Forward

   The Grandparent's new Guide to the Universe, *by Jaylynn Loreman and Nancy Woodard has produced a masterpiece of facts lavishly illustrated with tons of pictures. This book is definitely a must-have for Grandparents everywhere. Dan Milburn, (Grandpa) is just the helper for this" labor **of love**"* he calls it. The chance to do this project for Grandparents and Grandchildren alike, with my Grandchildren, is a monument to Grandparents and Grandchildren everywhere. It is a testimonial to tea-parties, wearing funny hats and building memories that will last a lifetime. Happy reading – together.

Other books you may be interested in:

I0500254

The
Grandparent's New Guide to the Universe

`The *new* Grandparent's Guide to the Universe is available both in book form and as a downloadable E-Book from Kindle Books at Amazon.com. It is lavishly illustrated with tons of pictures; however, when you crop pictures sometimes they are hard to see so here is a website where they can be seen full size if you want. http://facebook.com/better.image.1

The chapters ahead are stand-alone chapters, unlike in a talking book, they are each full article is packed with information and they don't have to be read in a sequence to make sense.

The table of contents will help you see highlights of their contents. You can move around according to your needs. The whole point of this book is to help Grandparents gain easy access to tons of useful information that has been truth-checked, verified and field tested on lots of Grandkids.

It isn't perfect but its real close. It is my fondest wish that what you learn from this book will help you to answer tons of questions by adding the only missing but important ingredient, <u>you</u>.

Table of Contents

Chapter One   Teaching our Grandchildren
how to think not _what_ to think!

If I had a star for every reason I love my grandchildren, I would have the whole night sky.

Every grandparent or parent for that matter has at one time or another probably *cringed* when they anticipated what questions their grandchildren or children might someday be asking them. The reason why we so dread those questions is because we are not all that sure we will know the answers – let alone the right answers. Being a parent or a Grandparent puts us in a vicarious position because our answers can have an everlasting influence on the people we love. How they envision their world and fellow human beings and how they may act in a given situation in their future may very well depend on our answers. The fact that we are a loved and trusted source may mean the difference between our answer and an answer from somewhere else.

The point of writing this book was to put at your fingertips a collection of facts organized in such a way that will enable you to answer many of those questions. And for the ones you can't answer, hopefully in this book you will have the tools you need to seek out and find them.

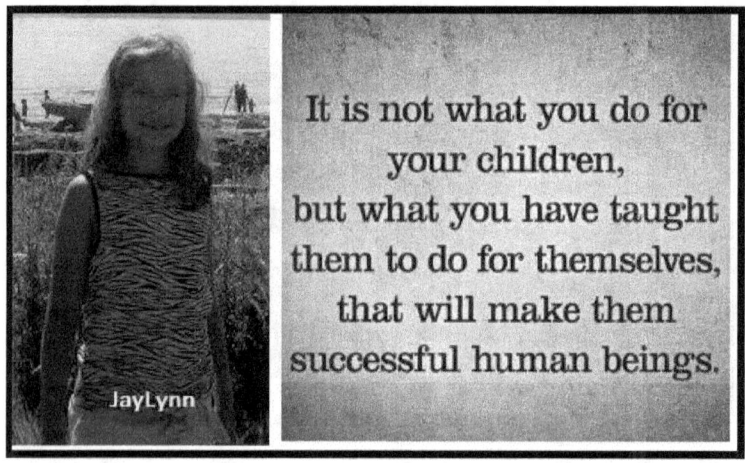

It is not what you do for your children, but what you have taught them to do for themselves, that will make them successful human beings.

JayLynn

JayLynn Loreman, (12) my granddaughter is also co-author of this book!

The moment that caused me to do a few mental *flip flops* was when my 4th grade granddaughter Jaylynn, said to me that in Sunday school she learned that God created her world in 6-days but in her science, class her teacher said that it took billions of years to create her world. At just ten years old she was facing what could be the greatest dilemma of her entire life.

I knew that her question had been answered hundreds of times before by some of the most respected scholars, priests, rabbis and scientists the world has ever known. Sometimes the answers were brief and concise, based on faith, philosophy or science, and other times the answers were based on complex formulas in physics. Some have been proven, some are still works in progress but one thing they all have in common is that each one of them has their own long list of followers and their own long list of critics.

The bathroom trick

I reacted to her question like any good grandfather might – I said, *"Hold that thought, I have to go to the bathroom – I'll be right back."*

While hiding in the bathroom I found myself in a quandary regarding what to teach my grandchildren about the origin of their world. On the one hand, I wanted them to know that a loving protective God watches over them whether they have been naughty or nice, and on the other hand that they live in an ever-expanding universe that scientists believe originated about 13.7 billion years ago.

Reality hit me

In the bathroom, I quickly stared at my shelf of reading material that included, magazines and few books etc. and then reality hit me. I had zip. I had nothing. Not even the internet. I started to sweat because to me being a grandfather is the highest achievement of my entire life and here I am after years of watching Disney Channel, spoiling them with candy and surprising them with *road trips* to the beaches, lakes and bouncy houses – I was about to let them down because I was about to lose all of that trust collateral I had built up with them – by failing the wisdom test. I could just imagine them talking to each and one of them says *"let's ask grandpa,"* and the other one says – *"why? He don't know."*

Suddenly I heard a voice in my head – my grandpa inner voice said - *"I have an idea, if nothing else it will buy me some time!"* I regained my composure as best I could and went back in the living room and said to my two granddaughters *"I have an idea and it's kind of a surprise so play along, ok? Grab your shoes and head for my truck."* We stopped and grabbed a hot pizza on the way to the beach, found a table and I said again – *"I have an idea"* - they said *"what?"* I said *"let's write a book. Let's make a list of the most important questions we can think of – and then let's search for the answers. We will discuss those questions and answers and we will write them down and make it all into a book."* Nancy, my younger granddaughter of the two suggested we develop folders with labels and hinted she needed a laptop computer for her new job! I said, *"I agreed if they agreed!"*

They quickly agreed. Like all kids they have a short attention span – especially when the sun is shining and they see other kids playing.

So, our first meeting of the minds was a success in the fact that we agreed on our next steps. As they ran off to explore the beach I let out a deep breath because I had dodged the bullet this time – but I knew it was temporary and that now that I was retired I was probably starting the toughest job in my life.

For the next two years the more we delved into researching both the Biblical and scientific information we realized that the two subjects seldom coincided peaceably. In fact, the level of hostility between the two groups of scholars is surprising.

The journey to gather information for this book at times resulted in mind blowing questions that came from the hearts and minds of my grandchildren and their friends.

If you've ever been stuck in the hot seat by your Grandkids then I'm sure you are familiar with the drill. One kid asks a question like *"What is God's name?"* And then that motivates the other kids to compete for your attention by asking another question. Any look of befuddlement or perplexity and suddenly they are rewarded for pushing your buttons and like a wolf pack they gather in a circle and begin attacking you with dozens more questions. Sometimes the questions are related to certain subject matter and sometimes they are from left field like *"what do you call Duck's feet?"* But rest assured, the aforementioned behavior is cemented into the instinctive learning behavior of kids everywhere and it is just a matter of time before it's your turn in the hot seat.

For instance, *"Grandpa, where did God hatch from?"* my 6-year old granddaughter Nancy asked after learning in Kindergarten that all living things are born or come from eggs.

Jaylynn, my fourth-grade granddaughter came home upset from school, saying: her teacher said, '*God didn't make people.*" He told her *"people came from monkeys."* She said she told him that *"maybe he came from a monkey but she sure didn't!"* *"Wasn't I right, Grandpa?"*

The simplest yet most challenging question it seemed came from 4th grader Jaylynn: *"Grandpa, do you believe in God?"*

And then later, I remember sitting there with my granddaughters after a close family friend died. One of them asked me what happens to people after they die.

I was perplexed to find that my usual grandparental truth compass was not so helpful in replying to these questions. For one, at the time I didn't know the truth myself. For another, I'm not sure if it really is all about truth: Is a 6-year-old capable of understanding intangibility? Can a 10-year-old accept randomness? Should I spare my 10-year-old granddaughter the journey I so laboriously traversed for over 60-years?

Moreover, I know those kids' questions about God can be driven by various motivations. While children often really do want to know, at other times they are seeking reassurance that their theory-building attempts to explain the world are acceptable. They may also ask about God out of a need for comfort, especially when stressed. While the first motivation may prompt sharing knowledge and the second listening, the third could involve bracketing one's personal truth.

But when my own children (their parents) came along many years ago and I found myself encouraging them to pray and anthropomorphizing God, I realized my own needs were at work too. For many doubters, the further we move away from childhood, the more we yearn for innocence and that unspoiled state of trust in the world. Since many of us have lost that option for ourselves, we may try to find it through our children. Sometimes we teach them things we don't believe in just because we want so badly to see that sweet innocence at work and experience unquestioning faith, if only by proxy.

I'll never forget how excited I was when Jaylynn lost her tooth. In my eagerness to provide the full experience, I wrote her a glitter-covered letter from the tooth fairy and put money in the envelope. After a while I began to wonder why on earth I was lying to my granddaughter. It seemed an almost sacred duty to make her believe in something I knew well to be nonsense. The inevitable corollary: *"is God a sort of cultural tooth fairy that adults pass on to their children once they can no longer hold on to him themselves?"*

"If you are a spirit, how can you make things which people can touch?" Nancy asked. Uncertain where to aim my answers, I offered none, instead describing ways in which people throughout history have attempted to handle these questions. The children didn't seem to mind the lack of definitive answers at all. It was asking the questions that mattered. I realized that by relinquishing the need to answer all children's questions about God, even the most conflicted doubter can discuss God meaningfully. And some bonus benefits can crop up, too.

Jaylynn recently said to me, *"I'd like to ask God if in the whole world there is someone who is only good and if in the whole world there is anybody who is only bad."* I told her that was a beautiful question, and then realized I had an answer. *"I don't think there is, I told her. The first – that no one is wholly good – makes us human. The second – that no one is wholly bad – makes us Godly."*

In the chapters, ahead you will find a lot of information about man's history that describe some of life's many secrets in an easy to understand language. Their purpose is to give us some necessary information that can help us understand which facts will help us to be a better-informed Grandparent. Trust me, you will thank me later.

JayLynn and cousin Nancy

Remember: Time waits for no one.

My granddaughters are now a few years older and hopefully much more capable of understanding the World around them and more importantly they are learning *how* to learn not *what* to learn.

History has been gathering written and spoken information for well over 5000-years on just about every subject imaginable. Libraries, cave drawings, sacred manuscripts, books and the internet are all within our grasp now. Seemingly one could have very little excuse for not knowing many of life's secrets, however; the sheer mountains of materials make access to all of it nearly impossible, unless you have a *way* to sort through it.

Have you ever wondered how we know the things that we know? How do we know, for instance, that the stars, which look like tiny pinpricks in the sky, are really huge balls of fire like the Sun and very far away? And how do we know that the Earth is a smaller ball whirling round one of those stars, the Sun?

The answer to these questions is **evidence**.

Sometimes **evidence** means actually seeing (or hearing, feeling, smelling….) that something is true. Astronauts have traveled far enough from the Earth to see with their own eyes that it is round. Sometimes our eyes need help.

The '*evening star*' looks like a bright twinkle in the sky but with a telescope you can see that it is a beautiful ball – the planet we call Venus. Something that you learn by direct seeing (or hearing or feeling…) is called an **observation.**

Often **evidence** isn't just **observation** on its own, but **observation** always lies at the back of it. If there's been a murder, often nobody (except the murderer and the dead person!) actually **observed** it. But detectives can gather together lots of other **observations** which may all point towards a particular suspect. If a person's fingerprints match those found on a dagger, this is **evidence** that he touched it. It doesn't prove that he did the murder, but it can help when it's joined up with lots of other **evidence**. Sometimes a detective can think about a whole lot of **observations** and suddenly realize that they all fall into place and make sense if so-and-so did the murder.

Scientists – the specialists in discovering what is true about the world and the universe – often work like detectives. They make a guess (**called a hypothesis**) about what might be true. They then say to themselves: if that were really true, we ought to see so-and-so. This is called a **prediction**. For example, if the world is really round, we can **predict** that a traveler, going on and on in the same direction, should eventually find himself back where he started. When a doctor says that you have measles he doesn't take one look at you and see measles. His first look gives him a **hypothesis** that you may have measles. Then he says to himself: if she really has measles, I ought to see… Then he runs through his list of **predictions** and tests them with his eyes (have you got spots?), his hands (is your forehead hot?), and his ears (does your chest wheeze in a measly way?). Only then does he make his decision and say, '*I diagnose that the child has measles.*' Sometimes doctors need to do other tests like blood tests or X-rays, which help their eyes, hands and ears to make **observations.**

Three bad reasons!

The way scientists use **evidence** to learn about the world is much cleverer and more complicated than I can write in this book. But now I want to move on from **evidence**, which is a good reason for believing something, and warn you against three **bad** reasons for believing anything. They are called (1) **tradition,** (2) **authority,** and (3) **revelation.**

First, **tradition:** A few months ago, I watched on television a man hosting a discussion with about 50 children. These children were invited because they'd been brought up in lots of different religions. Some had been brought up as Christians, others as Jews, Muslims, Hindus, and Sikhs. The man with the microphone went from child to child, asking them what they believed. What they said shows up exactly what I mean by '**tradition**'. Their beliefs turned out to have no connection with **evidence.** They just trotted out the beliefs of their parents and grandparents, which, in turn, were not based upon **evidence** either. They said things like, "*We Hindus believe so and so.*" "*We Muslims believe such and such.*" "*We Christians believe something else.*" Of course, since they all believed different things, they couldn't all be right. The man with the microphone seemed to think this quite proper, and he didn't even try to get them to argue out their differences with each

18

other. But that isn't the point I want to make. I simply would want to ask where their beliefs came from. They came from **tradition.** **Tradition** means beliefs handed down from grandparent to parent to child, and so on or from books handed down through the centuries. **Traditional beliefs** often start from almost nothing; perhaps somebody just makes them up originally, like the stories about Thor and Zeus. But after they've been handed down over some centuries, the mere fact that they are so old makes them seem special. People believe things simply because people have believed the same thing over centuries. That's **tradition.**

The trouble with **tradition** is that, no matter how long ago a story was made up, it is still exactly as true or untrue as the original story was. If you make up a story that isn't true, handing it down over any number of centuries doesn't make it any truer!

Most people in England have been baptized into the Church of England, but this is only one of many branches of the Christian religion. There are other branches such as the Russian Orthodox, the Roman Catholic and the Methodist churches. They all believe different things. The Jewish religion and the Muslim religion are a bit more different still; and there are different kinds of Jews and of Muslims. People who believe even slightly different things from each other often go to war over their disagreements. So, you might think that they must have some pretty good reasons – **evidence** – for believing what they believe. But actually, their different beliefs are entirely due to different **traditions.**

Let's talk about one particular **tradition.** Roman Catholics believe that Mary, the mother of Jesus, was so special that she didn't die but was lifted bodily into Heaven. Other Christian traditions disagree, saying that Mary did die like anybody else. These other religions don't talk about her much and, unlike Roman Catholics; they don't call her the *'Queen of Heaven'*. The **tradition** that Mary's body was lifted into Heaven is not a very old one. The Bible says nothing about how or when she died; in fact, the poor woman is scarcely mentioned in the Bible at all. The belief that her body was lifted into Heaven wasn't invented until about six centuries after Jesus's time. At first it was just made up, in the same way as any story like Snow White was made up. But, over the centuries, it grew into a tradition and people started to take it seriously simply because the story had been handed down over so many generations. The older the **tradition** became; the more people took it seriously. It finally was written down as an official Roman Catholic belief only very recently, in 1950, the year I was born. But the story was no truer in 1950 than it was when it was first invented 600 years after Mary's death.

I'll come back to **tradition** and look at it in another way. But first I must deal with the two other bad reasons for believing in anything: **authority** and **revelation.**

Authority, as a reason for believing something, means believing it because you are told to believe it by somebody important. In the Roman Catholic Church, the Pope is the most important person, and people believe he must be right just because he is the Pope. In one branch of the Muslim religion, the important people are old men with beards called Ayatollahs. Lots of young Muslims are prepared to commit murder, purely because the Ayatollahs in a faraway country tell them to.

When I say that it was only in 1950 that Roman Catholics were finally told that they had to believe that Mary's body shot off to Heaven, what I mean is that in 1950 the Pope told people that they had to believe it. That was it. The Pope said it was true, so it had to be true! Now, probably some of the things that Pope said in his life were true and some were not true. There is no good reason why, just because he was the Pope, you should believe everything he said, any more than you believe everything that lots of other people say. The present Pope has ordered his followers not to limit the number of babies they have. If people follow his authority as slavishly as he would wish, the results could be¹ terrible famines, diseases and wars, caused by overcrowding.

Of course, even in science, sometimes we haven't seen the **evidence** ourselves and we have to take somebody else's word for it. I haven't with my own eyes, seen the evidence that light travels at a speed of 186,000 miles per second. Instead, I believe books that tell me the speed of light. This looks like "**authority**" but actually it is much better than **authority** because the people who wrote the books have seen the **evidence** and anyone is free to look carefully at the **evidence** whenever they want. That is very comforting. But not even the priests claim that there is any **evidence** for their story about Mary's body zooming off to Heaven.

The third kind of bad reason for believing anything is called "**revelation.**" If you had asked the Pope in 1950 how he knew that Mary's body disappeared into Heaven, he would probably have said that it had been "**revealed**" to him. He shut himself in his room and prayed for guidance. He thought and thought, all by himself, and he became more and surer inside himself. When religious people just have a feeling inside themselves that something must be true, even though there is no **evidence** that it is true, they call their feeling "**revelation**." It isn't only popes who claim to have **revelations**. Lots of religious people do. It is one of their main reasons for believing the things that they do believe. But is it a good reason?

Suppose I told you that your dog was dead. You'd be very upset, and you'd probably say, 'Are you sure? How do you know? How did it happen?' Now suppose I answered: 'I don't actually know that Bella is dead. I have no **evidence**. I just have this funny feeling deep inside me that she is dead.' You'd be pretty cross with me for scaring you, because you'd know that an inside *'feeling'* on its own is not a good reason for believing that something is dead. You need **evidence**. We all have inside feelings from time to time, and sometimes they turn out to be right and sometimes they don't. Anyway, different people have opposite feelings, so how we to decide whose feeling are is right? The only way to be sure that a dog is dead is to see her dead, or hear that her heart has stopped; or be told by somebody who has seen or heard some real **evidence** that she is dead.

People sometimes say that you must believe in feelings deep inside, otherwise you'd never be confident of things like "*Grandpa loves me.*"

But this is a bad argument. There can be plenty of evidence that somebody loves you. All through the day when you are with somebody who loves you, you see and hear lots of little tidbits of evidence, and they all add up. It isn't purely inside feeling, like the feeling that priests call **revelation**. There are outside things to back up the inside feeling: looks in the eye, tender notes in the voice, little favors and kindnesses; this is all real **evidence**.

Sometimes people have a strong inside feeling that somebody loves them when it is not based upon any **evidence**, and then they are likely to be completely wrong. There are people with a strong inside feeling that a famous film star loves them, when really the film star hasn't even met them. People like that are ill in their minds. Inside feelings must be backed up by **evidence;** otherwise you just can't trust them.

Inside feelings are valuable in science too, but only for giving you, ideas that you later test by looking for **evidence**. A scientist can have a '*hunch*' about an idea that just feels right. In itself, this is not a good reason for believing something. But it can be a good reason for spending some time doing a particular experiment, or looking in a particular way for **evidence**. Scientists use inside feelings all the time to get ideas. But they are not worth anything until they are supported by **evidence.**

I promised that I'd come back to **tradition**, and look at it in another way. I want to try to explain why **tradition** is so important to us. All animals are built (by the process called evolution) to survive in the normal place in which their kind live. Lions are built to be good at surviving on the plains of Africa. Crayfish are built to be good at surviving in fresh water, while lobsters are built to be good at surviving in the salt sea. People are animals too, and we are built to be good at surviving in a world full of ... other people. Most of us don't hunt for our own food like lions or lobsters; we buy it from other people who have bought it from yet other people. We '*swim*' through a '*sea of people*'. Just as a fish needs gills to survive in water, people need brains that make them able to deal with other people. Just as the sea is full of salt water, the sea of people is full of difficult things to learn. Like language.

You speak English but your friend speaks German. You each speak the language that fits you to '*swim about*' in your own separate '*People Sea*'. Language is passed down by tradition. There is no other way. In England, *Bella* is a dog. In Germany, she is *ein Hund*. Neither of these words is more correct, or truer than the other. Both are simply handed down. In order to be good at '*swimming about in their people sea*', children have to learn the language of their own country, and lots of other things about their own people; and this means that they have to absorb, like blotting paper, an enormous amount of traditional information. (Remember that **traditional** information just means things that are handed down from grandparents to parents to children.) The child's brain has to be a sucker for **traditional** information. And the child can't be expected to sort out good and useful **traditional** information, like the words of a language, from bad or silly traditional information, like believing in witches and devils and ever-living virgins.

It's a pity, but it can't help being the case, that because children have to be suckers for **traditional** information, they are likely to believe anything the grown-ups tell them, whether true or false, right or wrong. Lots of what grown-ups tell them is true and based on **evidence** or at least sensible. But if some of it is false, silly or even wicked, there is nothing to stop the children believing that too. Now, when the children grow up, what do they do? Well, of course, they tell it to the next generation of children. So, once something gets itself strongly believed – even if it's completely untrue and there never was any reason to believe it in the first place – it can go on forever.

Could this be what happened with religions? Belief that there is a god or gods, belief in Heaven, belief that Mary never died, belief that Jesus never had a human father, belief that prayers are answered, belief that wine turns into blood – not one of these beliefs is backed up by any good evidence. Yet millions of people believe them. Perhaps this is because they were told to believe them when they were young enough to believe anything.

Millions of other people believe quite different things, because they were told different things when they were children. Muslim children are told different things from Christian children, and both grow up utterly convinced that they are right and the others are wrong. Even within Christians, Roman Catholics believe different things from

Church of England people or Episcopalians, Shakers or Quakers, Mormons or Holy Rollers, and all are utterly convinced that they are right and the others are wrong. They believe different things for exactly the same kind of reason as you speak English and someone speaks German.

Both languages are, in their own country, the right language to speak. But it can't be true that different religions are right in their own countries, because different religions claim that opposite things are true. Mary can't be alive in the Catholic Republic but dead in Protestant Northern Ireland.

What can we do about all this? It is not easy for you to do anything, because you are only human. But you could try this. Next time somebody tells you something that sounds important, think to yourself: *"Is this the kind of thing that people probably know because of **evidence**? Or is it the kind of thing that people only believe because of **tradition, authority** or **revelation?**"* And, next time somebody tells you that something is true, why not say to them: *"What kind of **evidence** is there for that?"* And if they can't give you a good answer, I hope you'll think very carefully before you believe a word they say.

In the chapters, ahead we have assembled many accurate portrayals of significant events in mankind's history that have previously shaped not only **what** we know but **how** and **why** we know it. When we view this information now as **evidence,** suddenly many *older* events and information begin to take on whole *new* and different meanings. In fact, we find that there is a lot of things we have been taught or have been teaching that are simply wrong or not true because they have been either passed down by **tradition,** revealed through someone's **revelation,** or directed by somebody's **authority** – yet most of it is not based on any real **evidence.**

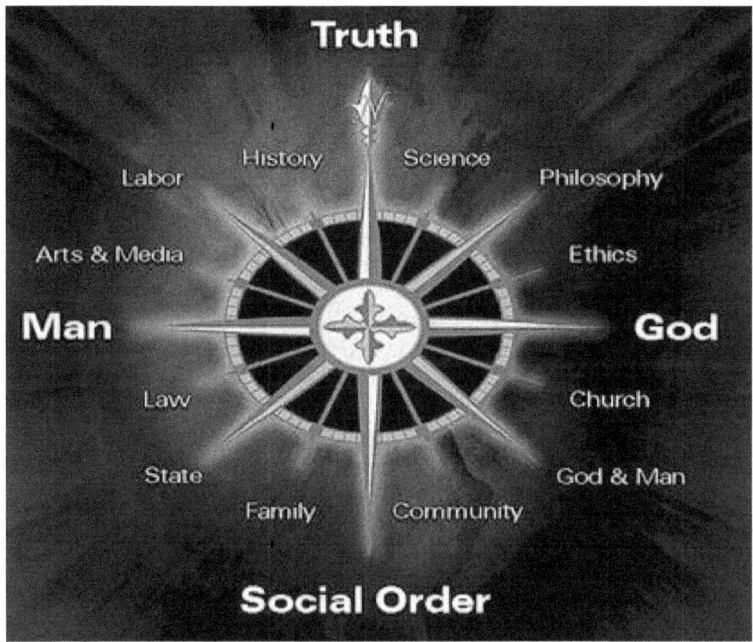

In today's civilizations, we find ourselves living in an information age like no other. We are bombarded from multiple sources with every subject imaginable containing everything from facts and truth to advertisements and intentional media misinformation. It is easily understandable how our individual truth compasses can become overloaded.

Whether we are grandparents or grandchildren our minds are capable of processing vastly more information than we need - it also is true that our minds function even much better if we weed out unreliable information – and even better yet - if we can weed it out **before** we learn it.

Hopefully we can teach our grandchildren that it's more important to learn *how* to learn rather than *what* to learn.

Preview:

Ahead in **Chapter two,** *"The Birth of Civilizations,"* you will notice that there is great attention paid to the period in time about 5000-years ago. That is because – as you will learn – overwhelming evidence shows disease wiped out all but a few breeding pairs of people on Earth. These surviving people were your ancestors. The proof is in your DNA. As they survived they also evolved into smarter human beings developing farming methods, raising animals, making advances in architecture, education and establishing early civilizations. Today virtually every man and woman on Earth can trace their ancestry back to these very same people. Helpful information to know whenever your grandchildren start asking questions about *"where did we all come from – and how did we get here."*

In **Chapter Three**, *"Angels and Ancestors,"* we look at historical records that show many of our ancestors felt the overpowering need to have spiritual connections. The need to believe in God or "a" God is evidenced in countless examples throughout the history of religions. And it is from that long history of all those different beliefs and religions - that religions are based on in today's world. This too is good information to have because in terms of causes and effects, it explains many *"whys"* in man's common human nature.

In the **Fourth Chapter**, through Jewish traditions, we are introduced to Abraham and we follow his life from the time he challenged his father's choice to worship and sell idols – to his revelation of a covenant with God that if he traveled to the land we know now as Israel – that God would give the title to that land to Abraham and his descendants. For the next thousand years Abraham and his descendants are said to have fathered not only the Jewish nation but the Arabs as well. During this time, they began construction of not only Jewish civilization but the Hebrew Scriptures as well. This is as good as any place in history to describe the Birth of the first Bible in the world.

In **Chapter Five,** historical evidence, traditions and authority are all intertwined detailing over 40-years of the Jewish attempt to construct and layout the perfect blueprint for Jewish life and Israel's future. 120 of the most intelligent and powerful Jews from all over Israel came together to form *"The Great Assembly." For over forty-years they laboriously rewrote and edited the "Hebrew Scriptures,"* to establish new rules for prayer, education, health and welfare of their people and Jewish law. Their final act was to seal the Hebrew Scriptures for all time.

The Jews of Israel have the most well documented history of any nation on Earth. They, to many people reflect the birthplace of mankind and his relationship with God. This chapter not only contains a lot of useful information but a unique insight into the character of the Jewish people.

Chapter Six

The Greeks come to Israel Like a wrecking crew. They destroyed the Jewish Temples, scattered the population, packed up Solomon's Library and sent it off to Athens, Greece where they adopted his philosophy and wisdom as their own. They replaced the God of Israel with the many gods of their own like Zeus and Poseidon and began assimilating much of the Jewish culture into their own.

This chapter exposes the fact that Western Civilization got much of its knowledge and culture from the treasures that were looted from Israel by Alexander the Great. In Greek history, he is known simply as *The Great Greek.*

Chapter Seven

The Greeks are still occupying Israel and the Jews have no armies or even the will to cast them out. In a final act of their aggression they replace the Jewish Scriptures with their own Greek version of what we now refer to as The Greek Bible, the Septuagint. The Greek Bible was formulated to promote Western civilization and became the key point of departure for countless versions of *The Bible* in the future.

Chapter Eight

From the Hebrew Scriptures, the Greek Bible (Septuagint) through many translations and versions filled history – and the Bible as we know it today in its many versions has been an ongoing work in progress which continues even today. Over the years the Bible has been rewritten and re translated to adapt to the ever-changing social values of its readers. Many famous quotes are as different as night and day in many of the new versions and today entire chapters are either missing or have been added.

A Brief History of the Bible and its many versions
Chapter Nine

Not every religion feels or views **faith** in the same way. Now there are as many views as there are religions yet still they all have many common threads. In today's world where we can travel easily it seems important that we try to understand some of those common threads as well as some of their differences. In some lands ignorance of their customs is viewed as no excuse and hostilities can easily develop - helpful information in an age where much of the Middle East is in flames because of it. In our world, today we are exposed to many people who live, dress and act differently because of their religion. This chapter gives us a glimpse of the people behind the head scarves and traditions and some of their beliefs. An interesting point is that we can be amazed at how truly similar we all are.

What other religions say about Faith

Chapter Ten answers many questions that occupy all our minds. Why do we pursue religious in the first place? Is God real or a figment of our imagination? Why is it so easy to convince so many people to join a religious group? Are we ready-made suckers for a pitch or is there something much more than that?

Truth is – yes and yes again, there really is much more to it than that. And what's more – science has the evidence to prove it!

What Science says about Faith?

In the previous ten chapters, there has been included a lot of information that many of us have seen or heard before, however; for the most part, our Grandchildren haven't. In this day and age, we can expect pieces of this information to be aimed at them daily – sometimes directly and other times by the way of friends, teachers or the media. Kids learn as much by just absorbing what appears to be going on around them as they do sitting at a desk. No lectures here. Just a reminder: It is not what we do for our children, but what we have taught them to do for themselves, that will make them successful human beings.

Within those previous ten chapters it becomes very much easier to understand how and why people believe what they do and don't about creation.

Chapter Eleven

The Birth of the Universe is fairly easy to explain in terms of the laws of physics and computing time and distances. In fact, as long as we stay in the areas of observable behaviors and predictable consequences it's a snap. It's when we spend too much time alone in a mountain top observatory with a calculator, pencils, paper and a deaf cat that information starts getting "iffy."

I hate to be the one to break it to some of you but contrary to popular belief there was really nothing very *"special"* about either the *"Big Bang,"* the *"Higgs Boson,"* or even the *"Birth of the Universe."* In fact, in terms of everyday science it was just another day in the Universe, business as usual.

The reason why so many of us perceive of those particular events as unusual or even *"special"* is because of our perception influenced by all the media hype and celebrating by discoverers who by the older more seasoned scientists are referred to as science newbies. Many are overzealous grant recipient's eager to justify spending millions on their projects. Throughout history mankind has continually made the mistake of believing that just because he *discovered* something that he *invented* it.

That being said, Chapter 11 gives us great information on how scientists are able to predict the age of the Universe and the processes that enabled the Earth to develop into a habitable planet. The formation of the atmosphere, the oceans and continents and even the moon are explained in pretty straight forward details. The evolution of living matter is detailed along with plenty of scientific charts and it becomes easy to imagine just how fragile our environment really is. The astronomically infinite number of variables that have occurred in order for there to be life on Earth are way too much for any computers as we know them to calculate. Unfortunately, the likelihood of another planet like Earth with life like ours on it is just as un-calculatable. Our takeaway is – simply put - when we answer the question *"Are we alone in the Universe?"* The answer is yep, probably maybe.........

Chapter Twelve

As in Chapter 11 the best scientists in the world using the most advanced technology that we have in our present day are busy exploring every inch of the Universe that they can either reach or see with a telescope still looking for clues of how we evolved, where we've been and where we are going.

All those numbers we talked about previously are presently scribbled on black boards all over the planet.

Why? Believe it or not they are still trying to answer an age-old question that even my preschool age granddaughter asked: "*Which came first – the chicken or the egg?*"

Similarly, at the end of most days, many of the world's top scientists are embroiled in the question of The Fine-Tuned Universe, "is the Universe fine-tuned for the development of life on Earth or is life on Earth fine tuned to the Universe and the laws of physics?"

Chapter Thirteen

The Quest for knowledge - What really happened to the dinosaurs? Today in a world full of environmental misinformation paid for by Giant Corporations scientific facts are often hard to come by. Many say that global warming and global cooling are simply normal stages that our planet goes through as they have for millions of years and how they will continue to do so. Others say that Global pollution and depletion of Earth's resources are to blame for unparalled extreme weather conditions throughout the planet and that eventually the planet is going to get tired of this nonsense and instead produce a toxic atmosphere to get rid of its biggest pest – mankind.

To point, many scientists believe that this is all just a rerun of what happened to the dinosaurs millions of years ago. That volcanos around the planet making up what geologists refer to as the ring of fire began erupting and filling the atmosphere with tons of ash and poisonous gases that completely blocked the sun forcing the Earth into millions of years of darkness. Dinosaurs had nothing to eat, no water to drink, no air to breathe and nowhere to hide and like much of the Earth's vegetation died off. Scientific discoveries in Geology and Paleontology confirm that fossil records worldwide confirm this catastrophic extinction of life on Earth.

In the last decade or so another group of scientists have been exploring yet another if not different theory about the extinction that then a likely event occurred simultaneously with the massive eruptions. And that is that a giant asteroid collided with the Earth with such force that it caused the Earth's atmosphere to become so full of debris that it too caused the Earth to fall into millions of years of darkness thus either causing directly or contributing significantly to the extinction of the Dinosaurs as well as most life on Earth.

Although science seems somewhat deadlocked concerning what killed off the Dinosaurs it appears a safe bet to consider that both events share responsibility.

At risk of appearing kind of scientifically naïve I am both puzzled by and intrigued not so much with the extinction of the Dinosaurs or even the catastrophic events leading probably to nearly destroying the planet – but by the fact that we all survived. Whether as a single cell ameba under a rock, an insect deep within the Earth, an amphibian under water or even the seed of a fern trapped in a fossil for a million years. I look up at the sky at all those stars, and I look down at the ground and see a seed germinating and most of all I look at my Grandchildren and imagine what their futures will be like. It is just overwhelming to consider the magnitude of the Earth's power to adapt and overcome and if that leaves any doubters then just ask yourself – how did we all survive? What will the Universe create next?

Chapter Fourteen *Fulfills our promise to focus on Grandkid's* questions and answers but goes a step further and lets them write the final chapter. The utter simplicity yet colorfully enlightening material is a testimonial to our

Grandchildren's endless supply of creativity. What started out as a project to teach my Grandchildren something turned out to be a valuable learning experience for me as well. With my greatest hopes, I wish for all of you many Grandparents to get involved in projects with your Grandchildren especially while they are still young enough to think we're still cool.

Perhaps the most important treasure that we can give our Grandchildren is to encourage their questions, to dwell in the light of their wonder at the many things we missed. I watched my granddaughter catch a raindrop on her finger and then hold it up to the sun and saw that it contained a rainbow inside. For that second in time nothing else in the entire Universe mattered. But inside her brain, captured for her eternity was the knowledge that there was a rainbow in every raindrop.

Chapter Two

The Birth of Civilizations - In the Beginning

In the beginning - around 5,000 years ago - the human race experienced a huge transition from knuckle dragging hunter gatherers to a more nomadic thoughtful group of people that became the forefathers of present day civilization. During this period, some of the people began traveling and exploring the vast new world around them discovering many new cultures, races and tribes. It became customary to bring new husbands and wives back to breed new blood into the tribes and to increase their influence and knowledge.

Unfortunately, it also had a profound negative impact on both those nomadic travelers and their hosts of other lands - it spread highly communicable diseases among them.

The main theory today is *"migration"* of peoples from other parts of the world may have been the cause of mass populations dying off. Simply put, each group of peoples carried *"diseases"* that although one group had developed immunity, to the newly introduced group the new diseases arrived and spread like a mass genocidal epidemic. The tiny group of surviving members of the population with acquired immunities to these newly acquired diseases survived and became our ancestors.

Both historical records and scientific evidence not only document this cataclysmic die off of nearly every person on Earth – but in a resounding dramatic way – in its aftermath also signals the Birth of civilization.

DNA studies

Scientists are in agreement and aware that in history while back there was an evolutionary bottle neck that reduced breeding pairs of humans down to a few.

Additionally, scientists studying DNA have also discovered that abrupt changes in brain chemistry occurred at the same time. However, depending upon the causes of the bottleneck, the survivors may have been those who were the most-fit individuals, hence improving the traits within the gene pool while shrinking it. Most genes in the genome are inherited from either father or mother, and thus can be traced back in time via either matrilineal or patrilineal ancestry. The ethnic and racial differences that play so large in today's societies were biologically meaningless then. Today only about 6,000 generations separate everyone alive from a common set of ancestors.

Written records of history

Written records of history began surfacing about 5,000 years ago. That puts the common ancestor of most all human beings alive approximately 5,000 years ago, or about 2985BCE. Scientists tell us that our ancestor probably lived somewhere in Eastern Asia, Taiwan, Malaysia or likely in Siberia. Those records seem to coincide with the advent of reading, writing and arithmetic. It also shows evidence of improved farming technics and technological advancements.

This ancestor lived and died doing nothing more remarkable than be born, live and have children. Yet this was the probable ancestor of most every human now living on Earth 5,000 years later. It is entirely possible that our ancestor was the last human in history whose family tree branches out to touch all 6.5 billion people on the planet today.

That means everybody on Earth descends from somebody who was around as recently as the reign of King Tut, (1332 – 1323 BCE) or aka Tutankhamen. There's even a chance that our last shared ancestor lived at the time of Christ in year 1 CE, or maybe even during the Golden Age of ancient Greece.

Scientists and historians have brought us much new information about **our ancestors** - who they were where they lived and what they were like. Lots of people now are tracing their genealogy and visiting historical sites hoping to capture just a glimpse of those who came before. Few people realize just how intricately these events connect them not just to people living on the planet today, but to everyone who ever lived.

With the help of statisticians, computer scientists and a supercomputer, science has calculated just how interconnected the human family tree is. You would have to go back in time only 2,000 to 5,000 years - and probably on the low side of that range - to find somebody who could count every person alive today as a descendant.

Everybody has the same set of ancestors

Additionally, now scientists are finding that if you go back a little farther - about 5,000 to 7,000 years ago - everybody living today has exactly the same set of ancestors.

Put in simpler terms, every person who was alive at that time is either an ancestor to all 6 billion-plus people living today or their ancestral line has died out or they have no remaining descendants.

That revelation is especially startling because had you entered any village on Earth in around 3,000 BCE the first person you would have met would probably be your ancestor.

It also means that all of us humans probably have ancestors of every race, color and nationality. It's all in the numbers. Every person has two parents, four grandparents and eight great-grandparents. Keep doubling back through the generations - 16, 32, 64, 128 and within a few hundred years you have thousands of ancestors. It's nothing more than exponential growth combined with the facts of life. By the 15th century you've got a million ancestors. By the 13th you've got a billion. Sometime around the 9th century - just 40 generations ago - the number tops a trillion.

The unavoidable question of "*how could anybody - much less everybody alive today have had a trillion ancestors living during the 9th century?*" The fact is - they didn't. Imagine there was a man living 1,200 years ago whose daughter was your mother's 36th great-grandmother, and whose son was your father's 36th great-grandfather. That would put him on two branches on your family tree, one on your mother's side and one on your father's.

In fact, most of the people who lived 1,200 years ago appear not twice, but thousands of times on our family trees, because there were only 200 million people on Earth back then. It's the math again - a trillion divided by 200 million - shows that on average each person back then would appear 5,000 times on the family tree of every single individual living today. Unfortunately, sometimes things are never average. Many of the people who were alive in the year 800 never had children; they don't appear on anybody's family tree. Meanwhile, more prolific members of society would show up many more than 5,000 times on a lot of people's trees.

As scientists keep going back in time, finding fewer and fewer ancestors available to put on the many branches of the 6.5 billion family trees of people living today - it becomes mathematically inevitable that at some point, there will be a person who appears at least once on everybody's tree.

Mind boggling as it seems, scientists keep going back and the number of potential ancestors dwindles and the number of branches multiply until there comes a time when every single person on Earth is an ancestor to all of us, except the ones who never had children or whose lines eventually died out. And it wasn't all that long ago.

When you walk through an **exhibit of Ancient Egyptian art** from the time of the pyramids, everything there was very likely created by one of your ancestors - every statue, every hieroglyph and every gold necklace. If there was a mummy lying in the center of the room, that person was almost certainly your ancestor, too.

In fact, it also means that when Muslims, Jews or Christians claim to be children of Abraham, they are all probably right.

No matter the languages we speak or the color of our skin, we share ancestors who planted rice on the banks of the Yangtze, who first domesticated horses on the steppes of the Ukraine, who hunted giant sloths in the forests of North and South America, and who labored to build the Great Pyramid of Khufu.

Proof in the numbers

Our researchers started thinking about how to estimate when the last common ancestor of everybody on Earth today lived. They discovered that there is a mathematical relationship between the size of a population and the number of generations back to a common ancestor. Plugging the planet's current population into this equation, they came up with just over 32 generations, or about 900 years.

Old answers were wrong

They knew that answer was wrong because it relied on some common, but inaccurate, assumptions that population geneticists often use to simplify difficult mathematical problems. For example, their analysis pretended that Earth's population has always been what it is today. It also assumed that individuals choose their mates randomly. And each generation had to reproduce all at once.

The researcher's calculations essentially treated the world like one big meeting place where any given guy was equally likely to pair up with any woman, whether she lived in the next village or halfway around the world. They were fully aware of the inaccuracy - people have to select their partners from the pool of individuals they have actually met, unless they are entering into an arranged marriage. But even then, they are much more likely to mate with partners who live nearby. And that means geography can't be ignored if you are going to determine the relatedness of the world's population.

The researchers knew they would have to account for geography to get a better picture of how the family tree converges as it reaches deeper into the past. They decided to build a massive computer simulation that would essentially re-enact the history of humanity as people were born, moved from one place to another, reproduced and died.

Our researchers created a program that put an initial population on a map of the world at some date in the past, ranging from 7,000 to 20,000 years ago.

Then the program allowed those initial inhabitants to go about their business. We allowed them to expand in number according to accepted estimates of past population growth, but had to cap the expansion at 55 million people due to computing limitations. Although unrealistic in some respects - 55 million is a lot less than the 6.5 billion people who actually live on Earth today - they found through trial and error that the limitation did not significantly change the outcome with regard to common ancestry.

Migration

The model also had to allow for migration based on what historians, anthropologists and archaeologists know about how frequently past populations moved both within and between continents. Our researchers chose a range of migration rates, from a low level where almost nobody left their native home to a much higher one where up to 20 percent of the population reproduced in a town other than the one where they were born, and one person in 400 moved to a foreign country.

Allowing very little migration, their simulation produced a date of about 5,000 BCE for humanity's most recent common ancestor. Assuming a higher, but still realistic, migration rate produced a shockingly recent date of around 1 CE Some people even suspect that the most recent common ancestor could have lived later than that.

Migration is the key here. When a people have offspring far from their birthplaces, they essentially introduce their entire family lines into their adopted populations, giving their immediate offspring and all who come after them a set of ancestors from far away.

People tend to think of preindustrial societies as places where this sort of thing rarely happened, where virtually everyone lived and died within a few miles of the place where they were born. But history is full of examples that belie that notion. Take Alexander the Great, who conquered every country between Greece and northern India, siring two sons along the way by Persian mothers.

MONGOLS, HUNS & VIKINGS

Alexander the Great

Or consider **Prince Abdi Al-Rahman**, son of a Syrian father and a Berber mother, who escaped Damascus after the overthrow of his family's dynasty and started a new one in Spain. The Vikings, the Mongols, and the Huns all traveled thousands of miles to burn, pillage and - most pertinent to genealogical considerations - rape more settled populations.

More peaceful people moved around as well. During the middle Ages, the Gypsies traveled in stages from northern India to Europe. In the New World, the Navaho moved from western Canada to their current home in the

South Pacific Islands

American southwest. People from East Asia fanned out into the **South Pacific Islands**, and Eskimos frequently traveled back

Eskimos traveled from Siberia to Alaska

and forth across the Bering Sea from Siberia to **Alaska.**

These genealogical networks, as they start spreading out have the ability to get virtually everywhere. Though people like to think of culture, language and religion as barriers between groups, history is full of religious conversions, intermarriages, illegitimate births and adoptions across those lines. Some historical times and places were especially active melting pots - medieval Spain, ancient Rome and the Egypt of the pharaohs, for example. And the point is you only need one.

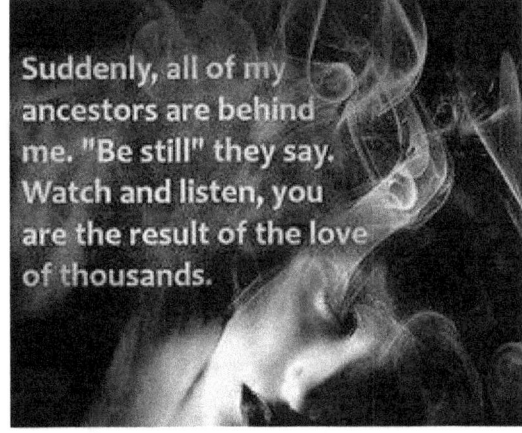

Suddenly, all of my ancestors are behind me. "Be still" they say. Watch and listen, you are the result of the love of thousands.

One ancestral link to another cultural group among your millions of forbears, and you share ancestors with everyone in that group.

So everyone who reproduced with somebody who was born far from their own native home - every sailor blown off course, every young man who set off to seek his fortune, every woman who left home with a trader from a foreign land - as long as they had children, they helped weave the tight web of brotherhood we all share. The lesson here is: ***this is where you came from!***

At this point, something else may be interesting for you to know. It is the fact that not only are you related to everybody else on Earth but you all have inherited many of the same instincts. *Chapter Three* will show what we are talking about.

Chapter Three

Where did Religion come from? When?

In a snapshot: In the last 4000-years or so there has been overwhelming **evidence** of mankind's numerous developments of religious beliefs. Religion is an organized system of beliefs and practices revolving around, or leading to, a transcendent spiritual experience.

There is **no** culture recorded in human history which has *not* practiced some form of religion. Every nation has created its own God in its own image and resemblance.

Religion, which, in ancient times, is indistinguishable from mythology concerns itself with the spiritual aspect of the human condition, gods and goddesses (or a single personal god or goddess), the creation of the world, a human being's place in the world, life after death and how to escape from suffering in this world or in the next.

The world's oldest religion still being practiced today is Hinduism (known to believers as *"Sanatan Dharma,"*

Eternal Order) but, in what is considered the *"West,"* the first records of religious practice come from Egypt around 4000 BCE.

This same pattern of creation of existence by a supernatural entity who speaks all into being, of how the world came to be as it is (the canopy of sky over the earth, for example) other supernatural beings emanating from the first and greatest one, a son who is a powerful entity himself who is killed or dies for his people and comes back to life for the good of his people and an afterlife similar to an earthly existence is **repeated** in religious texts from Phoenicia (2700 BCE) to Sumer (2100 BCE) to Palestine (1440 BCE) to Greece (800 BCE) and finally to Rome (100 CE).

The Phoenician tale of the great god Baal who dies and returns to life to battle the chaos of the god Yamm was already old in 2750 BCE when the city of Tyre was founded (according to Herodotus) and the Greek story of the dying and reviving god Adonis (600 BCE) was derived from earlier Phoenician tales based on Tammuz which was borrowed by the Sumerians (and later the Persians) in the famous Descent of Innana myth.

This theme of life-after-death and life coming from death and, of course, the judgment after death, gained greatest fame through the evangelical efforts of St. Paul who spread the word of the dying and reviving god Jesus Christ throughout ancient Palestine, Asia Minor, Greece and Rome (42-62 CE).

The religion of Christianity made standard a belief in an afterlife and set up an organized set of rituals by which a believer could gain everlasting life. In so doing, the early Christians were simply following in the footsteps of the Egyptians, the Sumerians, the Phoenicians and the Greeks all of whom had their own stylized rituals for the worship of their gods.

After the Christians, the Muslim interpreters of the Quran instituted their own rituals for understanding the supreme deity which, though vastly different in form from those of Christianity, Judaism or any of the older *'pagan'* religions, served the same purpose as the rituals once practiced in worship of the Egyptian goddess Hathor (3000 BCE) over five thousand years ago: to lend human beings the understanding that they are not alone in their struggles, suffering and triumphs, that they can restrain their baser urges and that death is not the end of existence.

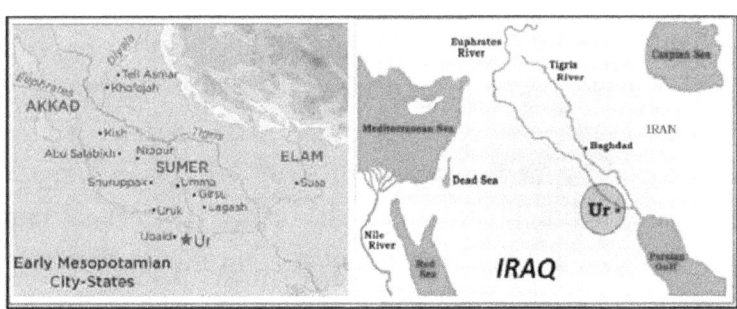

Scientific evidence, tells us that the development of cities is synonymous with **the rise of civilization**. Historical records tell us that early civilizations arose first in lower **Mesopotamia** (see above) around 3,500 BCE, followed by Egyptian civilization along the Nile around

3,000 BCE and the Harappan civilization in the Indus Valley in what we now know as Pakistan in about 2,500 BCE.

Among the earliest surviving _written_ **religious scriptures** are the Egyptian Pyramid Texts, the oldest of which date to between 2,400 and 2,300 BCE. Additionally, almost the entire written history of Israel and the Jews is based on the Hebrew Scriptures. By overlapping these historical records, we get a look at the world around 5,000 years ago which starts with the year 2,985BCE which was 2985 years before the Christian calendar would begin with the year 1CE. The expression of ideas became quite common with the advent of writing.

What were they like?

Idol Worship

Also, as complex civilizations arose, so did complex religions, and the first of their kind apparently originated during this period. Inanimate entities such as the Sun, Moon, Earth, sky, and sea were often deified. Shrines developed, which evolved into temple establishments, complete with a complex hierarchy of priests and priestesses and other functionaries. Typical of the Neolithic was a tendency to worship anthropomorphic deities or **idols.**

Who?

According to the Jewish scriptures it says that Abraham, then known as Abram, had a divine calling and left his home city and moved to the area of Canaan, now modern-day Israel.

Historical Jewish records indicate Abraham's father came from Mesopotamia (now Iraq) from a city called Ur.

Ur *(above)* is one of the oldest cities in the world and its ruins are still visible today. Ur is located on the edge of the Al-Hajar Desert in Iraq.

According to Jewish tradition, Abraham was born under the name Abram in the city of Ur in Babylonia in the year 1948 from Creation (circa 1800 BCE). He was the son of Terach, an idol merchant, but from his early childhood, he questioned the faith of his father and sought the truth. He came to believe that the entire universe was the work of a single Creator, and he began to teach this belief to others.

Abram tried to convince his father, **Terach**, of the folly of idol worship. One day, when Abram was left alone to mind

the store, he took a hammer and smashed all of the idols except the largest one. He placed the hammer in the hand of the largest idol. When his father returned and asked what happened, Abram said, *"The idols got into a fight, and the big one smashed all the other ones."* His father said, *"Don't be ridiculous. These idols have no life or power. They can't do anything."* Abram replied, *"Then why do you worship them?"*

Eventually, (*according to tradition*) the one true Creator that Abram had worshipped called to him, and made him an offer: if Abram would leave his home and his family, then his God would make him a great nation and bless him.

Abram accepted this offer, and the b'rit (*covenant*) between
God and the Jewish people was established. (Gen. 12).

The idea of b'rit is fundamental to *traditional* Judaism:
we have a covenant, a contract, with God, which involves
rights and obligations on both sides. We have certain
obligations to God, and God has certain obligations to us.
The terms of this b'rit became more explicit over time, until
the time of the Giving of the Torah. The Torah is the first
of the three books that make up the Tanakh.

The Torah is *"the book of laws."*

Abraham

Abraham offered to sacrifice his son.

Abram was subjected to many tests of faith to prove his worthiness for this covenant. Leaving his home was one of these trials. In another

Abram, raised as a city-dweller, adopted a nomadic lifestyle, traveling through what is now the land of Israel for many years. God promised this land to Abram's descendants. Abram is referred to as a Hebrew (Ivri), possibly because he was descended from Eber or possibly because he came from the *"other side"* (eber) of the Euphrates River.

But **Abram** was concerned, because he had no children and he was growing old. Abram's beloved wife, Sarai, knew that she was past child-bearing years, so she offered her maidservant, Hagar, as a wife to Abram. This was a common practice in the region at the time. According to tradition, Hagar was a daughter of Pharaoh, given to Abram during his travels in Egypt. She bore Abram a son, Ishmael, who, according to both Muslim and Jewish tradition, is the ancestor of the Arabs. (Gen 16)

When Abram was 100 and Sarai 90, God promised Abram a son by Sarai. God changed Abram's name to Abraham (*father of many*), and Sarai's to Sarah (from "*my princess*" to "*princess*"). Sarah bore Abraham a son, Isaac (*in Hebrew, Yitzchak*), a name derived from the word "*laughter,*" expressing Abraham's joy at having a son in his old age. (Gen 17-18). Isaac was the ancestor of the Jewish people. Abraham died at the age of 175 (c. 1813 - c. 1638 BCE)

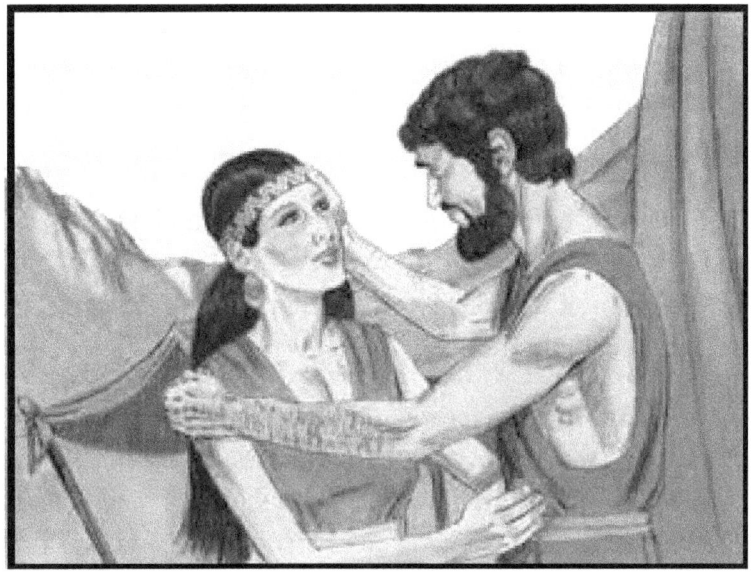

Isaac married Rebekah, who was very beautiful
and a distant relative. He also had twin sons, Jacob and
Esau. According to their Bible, Jacob lived up to the
popular translation of his name _"Jacob"_ figuratively means
"he deceives" and cheated Esau out of his inheritance as
the first-born. But Jacob was still the chosen one of God.
Legend says his second wife hired him out to his first wife
for a night of passion in return for a bunch of vegetables.
We can only presume that his life of hardship strengthened
him mentally and physically. Legend goes on to state that
after a wrestling match with God, his name was changed to
Israel, which means _"He struggles with God"_ - thus the
origin of the name, Israel. The Jewish people have, at
certain times, been known as the _**Children of Israel.**_ We
can look at this literally and discover that Jacob (_or
"Israel"_) had twelve sons.

Both the Arab and the Jewish societies developed a number of other unifying characteristics, including a central government, a complex economy and social structure, sophisticated language and writing systems along with distinct cultures and religions.

Scientific evidence, historical records and their Bible and all agree on one thing for sure - that when Muslims, Jews or Christians claim to be children of Abraham, they are all probably right.

In the Middle Eastern world - Jews have more than one definition. According to one definition Jews are not Jewish by birth, another definition states that they are - and yet still another states, that a Jew is any person of the Jewish faith. In today's Jewish world we find that many mixed races and cultures make up much of a society inhabited by people of all hues and mixtures of traditions.

Abraham, believed and taught all of his descendants that his God was the creator of the Universe!

Chapter Four

The first Bible in history was the Hebrew Bible and versions of it have been dated to over a thousand years. Though the word "*Bible*" is commonly used by non-Jews -- as are the terms "*Old Testament*" and "*New Testament*" -- the appropriate term to use for the Hebrew Scriptures is Tanakh.

Ta=Torah
Na=Nebium
Kh=Kethubim

The Hebrew Bible or Tanakh

The Hebrew Bible, or Tanakh, (*pronounced "ta-nock"*) contains twenty-four books divided into three parts; the first part includes the five books of the Torah that are for "*teaching*" and are considered religious "*law*", the second part is known as the Nevi'im, which is about the "*prophets,*" and the third part is called the Ketuvim or the "*writings*".

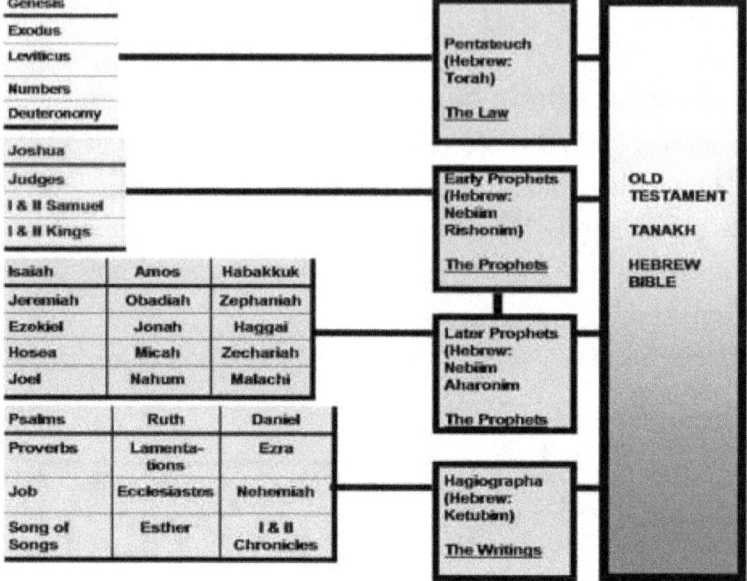

Development of the Hebrew Bible canon

Canon: refers to any standard or convention. The corresponding adjective is canonical. English canon may also represent Latin canonicus "*one who is canonical*."

According to rabbinic tradition, all of the teachings found in the Torah, both written and oral, were given by God to Moses, some of them at Mount Sinai and others at the Tabernacle, and all the teachings were written down by Moses, which resulted in the Torah we have today.

The Masoretic Text (MT) is the authoritative Hebrew text of the Jewish Bible. While the Masoretic Text defines the books of the Jewish canon, it also defines the precise letter-text of these biblical books, with their vocalization and accentuation known as the Masorah. The MT is also widely used as the basis for translations of the Old Testament in Protestant Bibles, and in recent years (since 1943) also for some Catholic Bibles, although the Eastern Orthodox continue to use the Septuagint, as they hold it to be divinely inspired. In modern times the Dead Sea Scrolls have shown the MT to be nearly identical to some texts of the Tanakh dating from 200 BCE but different from others. The oldest Hebrew (*Tanakh)* manuscript in both Hebrew and Aramaic dates to the 10th century CE. Tanakh is a Hebrew word: and reflects the threefold division of the Hebrew Scriptures, Torah "*Teaching*", Nevi'im "*Prophets*" and Ketuvim "*Writings*".

The oldest extant manuscripts of the Masoretic Text date from approximately the 9th century CE, and the Aleppo Codex once the oldest complete copy of the Masoretic Text, but now missing its Torah section) dates from the 10th century CE.

The **Torah** is also known as the "**Five Books of Moses**" or the **Pentateuch**, meaning "*five scroll-cases*". The Hebrew names of the books are derived from the first words in the respective texts. The Torah comprises the following five books:

- Genesis, Bereshith *"origin"*

- Exodus, Shemot *"going out"*

- Leviticus, Vayikra *"relating to the Levites"*

- Numbers, Bamidbar *"numbering of the Israelites"*

- Deuteronomy, Devarim *"second law"*

Jewish religious law

The Torah contains the commandments of God, revealed at Mount Sinai (although there is some debate among traditional scholars as to whether these were all written down at one time, or over a period of time during the 40 years of the wanderings in the desert, while several modern Jewish movements reject the idea of a literal revelation, and critical scholars believe that many of these laws developed later in Jewish history.

These commandments provide the basis for Jewish religious law.

The first eleven chapters of Genesis provide accounts of the creation (or ordering) of the world and the history of God's early relationship with humanity.

The remaining thirty-nine chapters of Genesis   provide an account of God's covenant with the - Biblical patriarchs Abraham, Isaac and Jacob (*also called Israel*) and Jacob's children, the "*Children of Israel*", especially Joseph.

It tells of how God commanded Abraham to leave his family and home in the city of Ur, eventually to settle in the land of Canaan, and how the Children of Israel later moved to Egypt.

The Giving of the Torah to Moses at Mt. Sainai.

The remaining four books of the Torah tell the story of **Moses**, who lived hundreds of years after the patriarchs. He leads the Children of Israel from slavery in ancient Egypt to the renewal of their covenant with God at Mount Sinai and their wanderings in the desert until a new generation was ready to enter the land of Canaan. The Torah ends with the death of Moses.

The Bible was written by around forty different people from different backgrounds, from kings, prophets, and writers to fishermen, shepherds, and prisoners. The Bible was written during a period of 1,600 years. That's about forty generations. The first division into chapters was made in 1238. The first division into verses was made in 1448 (*Tanakh*) and 1551 (*New Testament*). The first full Bible divided into chapters and verses was the Geneva Bible, printed in 1560.

The **Bible** was the first book ever printed in 1456. The Bible is the most sold and most translated book in the world. The Bible, or parts of it, is available in 2,508 different languages (UBS

(Figures of December 31, 2012). Every minute about fifty Bibles are sold.

The oldest preserved Bible is Codex Vaticanus, written before the year 350 AD. It is preserved in the Vatican Museum in Rome. The oldest preserved fragment of a Bible text is 2,600- *years* old. Orthodox and Catholic Christians include a number of additional books in their Old Testament (commonly referred to as the Apocrypha or Deutrocanon) and some additional material in books accepted by Protestants (Daniel and Esther).

The complete set of scrolls, constituting the entire Tanakh, is the name used in Judaism for the canon of the Hebrew Bible. Canons refer to authorized; recognized; accepted: canonical works.

The **Tanakh** is also known as the Masoretic Text or the Miqra. The name is an acronym formed from the initial Hebrew letters of the Masoretic Text's three traditional subdivisions: The Torah *(teaching)* also known as the Five Books of Moses, Nevi'im *(prophets)* and Ketuvim *(writings)* hence TaNaKh. The name Migra meaning that, which is read, is an alternative Hebrew term for the Tanakh. The books of the Tanakh were relayed with an accompanying oral tradition passed on by each generation, called the Oral Torah.

(Talmud: book of instructions written by thousands of Rabbis and is the basis of all Jewish Law.)

According to the *Talmud*, much of the contents of the *Tanakh* were compiled by the Men of the Great Assembly by 450 BCE, and have since remained unchanged. The Men of the Great Assembly was composed of the 120 greatest leaders, some of whom were prophets, including Ezra, Nehemiah, Mordechai, Zerubbabal, Haggai, Zechariah and Malachi. Modern scholars believe that the process of canonization of the Tanakh became finalized between 200 BCE and 200 CE.

The Hebrew text was originally an abjad: consonants written with some applied vowel letters (*matres lectionis*). During the early Middle Ages scholars known as the Masoretes created a single formalized System of vocalization. This was chiefly done by Aaron ben Moses ben Asher, in the Tiberias School, based on the oral tradition for reading the Tanakh, hence the name Tiberian vocalization. It also included some of Ben Naftali and Babylonian innovations.

Despite the comparatively late process of codification, some traditional sources and some Orthodox Jews believe the pronunciation and cantillation derive from the revelation at Sinai, since it is impossible to read the original text without pronunciations and cantillation pauses. The combination of a text (miqra), pronunciation (niqqud) and cantillation (te`amim) enable the reader to understand both the simple meaning, as well as the nuances in sentence flow of the text.

The cantillation signs are often an important aid in the interpretation of a passage. For example, the words qol qore bamidbar panu derekh YHWH (*Isaiah 40-3*) is translated in the Authorized Version as the voice of him that cried in the wilderness, Prepare ye the way of the LORD. As the word qore takes the high-level disjunctive zaqef qaton this meaning is discouraged by the cantillation marks. Accordingly, the New Revised Standard Version translates A voice cries out: '*in the wilderness prepare the way of the LORD*", while the New Jewish Publication Society version has A voice rings out: '*Clear in the desert a road for the LORD.'*

The three-part division reflected in the acronym "*Tanakh*" is well attested to in Rabbinic literature. During that period, however, "*Tanakh*" was not used. Instead, the proper title was **Mikra**, meaning *reading* or *that which is read* because the biblical texts were read publicly. Mikra continues to be used in Hebrew to this day, alongside Tanakh, to refer to the Hebrew Scriptures.

Remember: **The first Bible in history** was the **Hebrew** *Bible and versions of it have been dated to over a thousand years. Though the word "Bible" is commonly used by non-Jews -- as are the terms "**Old Testament**" and "**New Testament**" -- the 'appropriate' term to use for the Hebrew Scriptures is **Tanakh.***

Chapter Five

The Great Assembly

The Great Assembly

According to rabbinic tradition, all of the teachings found in the Torah, both written and oral, were given by God to Moses, some of them at Mount Sinai and others at the Tabernacle, and all the teachings were written down by Moses, which resulted in the Torah we have today.

Over the next 2500 years

Since then over a period of about 2500 years over 40- people have written and rewritten the Hebrew Scriptures.

It's worth remembering that according to the Talmud, much of the contents of the Tanakh were compiled by the Men of the Great Assembly by 450 BCE, and have since remained unchanged. The Men of the Great Assembly was composed of the 120 greatest leaders, some of whom were prophets, including Ezra, Nehemiah, Mordechai, Zerubbabal, Haggai, Zechariah and Malachi. Modern scholars believe that the process of canonization of the Tanakh became finalized between 200 BCE and 200 CE.

The Men of the Great Assembly passed decrees that ensured the Jewish peoples' survival in the post-Temple era down to our times.

At a time when Jewish life in the Land of Israel was crumbling, Ezra and Nehemiah swept in like a whirlwind.

At a time when Jewish life in the Land of Israel was crumbling, Ezra and Nehemiah swept in like a whirlwind. They not only closed the breaches in the physical walls of Jerusalem, built the Second Temple and set the foundation for the Second Commonwealth (the Second Temple era), but set the spiritual foundation and built the spiritual walls of the nation for the foreseeable future lasting to this day.

The primary vehicle through which they accomplished this was the establishment of an executive and legislative body called the Anshei Knesset HaGedolah, the "*Men of the Great Assembly*," which was composed of the 120 greatest leaders, some of whom were prophets, including Ezra, Nehemiah, Mordechai, Zerubbabal, Haggai, Zechariah and Malachi. The modern day Israeli Knesset (Knesset is the Hebrew word for "*Assembly*"), which has 120 members, took its name and number from them.

Unfortunately, for many Jews, that is where the comparison ends.

Their primary task was to smooth the transition into the new era by passing legislation that would insure the survival of the Jewish people.

Among the most important measures were sealing the scriptures, instituting prayer, coordinating the Jewish calendar and establishing an educational system in the land of Israel.

All told, the Men of the Great Assembly spanned no more than a generation. Yet, the reverberations of their decisions are still felt today. By studying their decrees, we gain entrance to a glimpse not only of their role in Jewish history but of the primary ingredients needed for Jewish survival.

First: Sealing the Biblical Canon

Their first major decision was to seal the Bible; to decide which books to include in the Holy Scriptures. Prophecy had ceased in the Second Temple era (Talmud, Sotah 48b). No longer could anyone claim to have an open line, as it were, with God. Therefore, without prophecy there was no possibility of divine inspiration to warrant admittance of anyone's words into the Bible.

One reason this was important was because later groups like the Christians felt it necessary to include certain books which are not included in the Jewish reckoning. These books, they wanted to believe, bridged the gap to the

Christian Gospels, a gap for all intents and purposes closed centuries beforehand with the decree to seal the Bible.

In all, the Men of the Great Assembly included 24 books in the counting. Afterward, no more books could be added. As the central body of the greatest Jewish sages and scholars their authority to do so was never disputed.

Second: Prayer

The second monumental accomplishment of the Men of the Great Assembly was the formulation of a universal Jewish prayer service.

Today, the centerpiece of every service is the prayer known as the Amidah (literally the *"standing" prayer*). It and its attendant prayers were apparently absent, in the First Temple era. The need for such a formalized prayer only first arose when the Jews went into exile in Babylon. The missing experience of community that went part and parcel with the three-times-a-year pilgrimage to the Temple left a vacuum. Without the Temple, essential nutrients in the peoples' religious diet were lacking. Therefore, the leaders in Babylon codified a system of prayer that substituted the Temple service. They based this on the prophetic verse, *"Our lips will substitute for sacrifices"* (Hosea 14:3).

When the Jews returned from Babylon to the Land of Israel and rebuilt the Temple they brought along with them the prayers they had learned in Babylon. The Men of the Great Assembly ordered, edited, and formulated the words of the Amidah, as well as its surrounding prayers. This arrangement continued through the entire Second Temple era and remains today.

Although the individual synagogue system was inferior, it successfully compensated for the shift in Jewish life away from the centralized Temple system. Now, with the stamp of approval from the Men of the Great Assembly, Jewish prayer became possible in each community, in each individual, no matter how far away he/she was. Instituting prayer this way not only substituted for the Temple service but compensated for the loss of center in Jewish life.

Third: The Calendar

Their third major accomplishment was the development of the permanent calendar. When the communists came to power in Russia in 1917 they banned the Jewish calendar even before they banned the prayer-book. They realized that without knowing the precise dates of the Jewish holy days no Jew could possibly maintain his religion. If one Jew thought Yom Kippur was Wednesday and one thought it was Thursday and another thought Friday the structure of Jewish life would collapse. Therefore, they banned the calendar first.

The Jewish calendar is based on the cycle of the moon. However, if it were a strict lunar calendar then every year would be 11¼ days less than the solar year. The problem then would be that in three years an entire month would be lost. In six years, two months would be lost. And so forth. Eventually, a holiday like Passover would come out in the dead of winter… and then fall… and then summer… and then spring again.

In fact, this is precisely the situation with the Muslim calendar, which is entirely lunar. After a period of time their holidays traverse the entire year. The Torah, however, expressly commands that Passover fall out in the springtime (Exodus 23:15, 34:18; Deuteronomy 16:1), which refers the vernal equinox.

Therefore, the Jewish sages added a leap month to the Jewish year. The solar and lunar years line up exactly every 19 years. Therefore, seven times every nineteen years an entire month is added.

Not only were the Jewish leader's great Torah scholars but they had great knowledge in astronomy, mathematics as well as other *"secular"* disciplines. Among the proofs of their intellectual acuity are the mathematical calculations they made back then for determining the exact moment of the lunar month. The Talmud (Rosh Hashanah 25a) tells us that, based on those calculations and a tradition going back to Sinai, Jewish months are calculated at 29.53059 days.

Only first with the advent of modern technology -solar satellites, hairline telescopes, laser beams and super-computers - were NASA scientists able to determine the length of the *"synodic month,"* i.e. the time between one new moon and the next. And that figure is 29.530588 days.

The basis for the permanent calendar was laid by the Men of the Great Assembly. They perfected all of the mathematical adjustments and intricacies that make the Jewish calendar laser-beam accurate even by today's standards — a truly remarkable feat.

Fourth: Education System

The fourth accomplishment of the Men of the Great Assembly was the establishment of an educational system in the Land of Israel.

From time, immemorial (*way back when*), Jewish survival has always depended on knowledgeable Jews. As we mentioned previously, the settlers who originally returned with Ezra and Nehemiah were not the elite of the Jewish people. Consequently, the educational system was at best mediocre. It was the task of the Men of the Great Assembly to revamp and revitalize it.

The problem they faced was not only raising funds, but building schools and attracting students. Israel had lost credibility as the Torah center for world Jewry. It was surrounded by enemies and economically unstable. They, therefore, mounted a campaign to reestablish Israel, and specifically Jerusalem, as the central address of the Jewish people.

One of the first things that they did toward that end was to re-establish the Sanhedrin, the central body of Torah authority. Once operational again they let it be known that all important questions of Jewish law should be sent to Jerusalem.

Ancient documents discovered about a century ago bear out how well this strategy worked. The documents referred became known as the Elephantine Papyri. The papyrus parchments were made from reeds grown along the shores of the Nile. They were written by the members of a garrison of Jewish mercenaries stationed near what is today the Aswan Dam. The Greeks called it Elephantine because they transported elephants to and from there.

This group consisted of about 150 soldiers, along with their families, who were on permanent station in Elephantine. It is mind boggling. Elephantine was the end of the world at that time. These 150 Jews were hired to collect customs duties for the Persian Empire and prevent marauding African tribes from entering Egypt. Some job for a nice Jewish boy!

The Elephantine Jews knew they were Jews, but little else. They wrote a letter to the High Priest in Jerusalem, asking him basic questions such as what day of the year was Rosh Hashanah, where do they get matzos (unleavened bread) for Passover, how do they make them, how do they build a mikvah (*bath for ritual immersion*), which way was the synagogue supposed to face, etc. In essence, the letter said, "*Make us Jews; we want to be Jews; help us.*"

Since these letters were written in Aramaic we can surmise that the people originated in Babylon. This was exactly what the Men of the Great Assembly worked to bring about — to reestablish the central address for all Jewry as Jerusalem.

Political Fallout

Even without necessarily making decrees, the mere strength of the Men of the Great Assembly brought about at least two very significant political results:

1) The absence of a monarch of Davidic ancestry.

2) The strengthening of the office of the High Priest.

As soon as Ezra established the Men of the Great Assembly no Davidic ancestor ever again played a prominent role on the national level. At the same time, power gradually shifted into the office of the High Priest.

In the First Temple era, the Davidic kings ran the country and the High Priest was completely subservient to them. By contrast, in the first 300 years of the Second Temple era, until Herod, the High Priest was inordinately powerful. He played a prime role in not only the religious life but the political, military, and diplomatic arena as well. Therefore, it is no surprise that the Maccabees, who were priests, eventually saw themselves fit to take the reign of Jewish kingship.

Misuse of power eventually plagued the office of High Priest. However, for about 150 years it remained uncorrupted. As long as men such as Simon the Just (*the High Priest of his time*) wielded influence, the Second Commonwealth continued to gain momentum.

When Simon died it seemed the sun slowly set on an era of sound leadership. Corruption crept in afterward with the Jewish assimilation, the establishment of the non-Davidic monarchy and the coming of the Greeks.

Chapter Six

The Greeks come to Israel

Greek culture would tempt Jews like no other, threatening to destroy Judaism. It would be as much an exile of mind as of the body. Jewish history does not happen in a vacuum.

Jews may play a disproportionate role in world history, but that role must be understood against the backdrop of the history of the entire world. In a deeper sense, Jews see it as God's hidden hand becomes visible in the behavior of the world toward the Jews.

How Jews react to that hand is very much a part of their overall plan.

Birth of Western Civilization

The rise and spread of Greek civilization affected the course of the non-Jewish world as perhaps no other historical force. Greek culture and philosophy formed the foundation for much of what today is known as Western civilization. The arrival of the Greeks as a prominent power in the Mediterranean basin greatly affected the Jewish settlement in Israel as well.

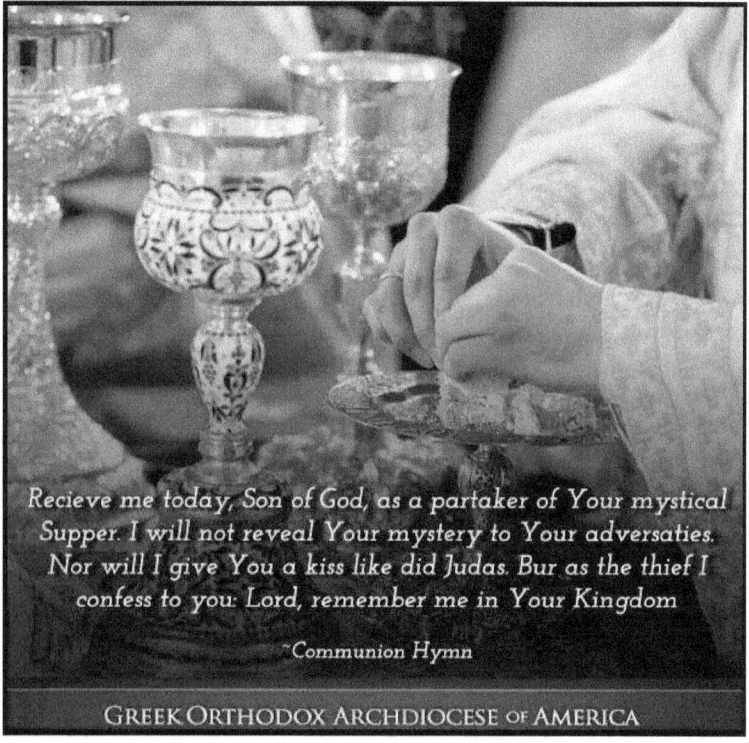

Recieve me today, Son of God, as a partaker of Your mystical Supper. I will not reveal Your mystery to Your adversaties. Nor will I give You a kiss like did Judas. Bur as the thief I confess to you: Lord, remember me in Your Kingdom

~Communion Hymn

GREEK ORTHODOX ARCHDIOCESE OF AMERICA

More overwhelming than the political or military threat which the Greeks posed was the spiritual threat. Jewish allegiance to the Torah was challenged as at no other time.

Greek values often clashed with the Jewish ideal, and, therefore, the infiltration of Greek culture and ideas set the stage for one of the most intriguing chapters in Jewish history.

The #1 Rule about Assimilation

For Jews today, more immediate than the historical events of this period is the establishment of a pattern for Jewish assimilation, a historical pattern which remains applicable to the Jewish people throughout the ages.

The essence of the idea is this: Jews <u>never</u> attempt to assimilate into an *inferior* culture.

The pressure for assimilation always exists and is strong when there is what the Jews consider an equal or superior culture involved. Therefore, the Jews did not assimilate in Babylon because they viewed it as an *inferior* culture. There was no temptation. Similarly, the Jews were never really interested in being Persians. The Jew in Eastern Europe did not want to be the Polish peasant. He did not want to be the Russian serf.

However, the Jew who lived in Germany wanted to be Schiller or Goethe or Frederick the Great. There was an attraction, because the Jews considered it an *advanced* culture.

The first major assimilatory threat to the Jewish people was Greek culture. For the first time, the Jews not only encountered a culture that provided an alternative, but, on the surface at least, provided a superior culture. That is why there grew such a great and strong Hellenistic movement within the Jewish people.

Greek Philosophy

Probably the most famous aspect of that culture is Greek philosophy.

It is an oversimplification, but the purpose of philosophy is to try to explain life logically. As such, it is like sleeping in a bed with a blanket that is a little too short. Something is always sticking out. There has never been a philosophy that answers all the questions.

In our time, the value of philosophy has declined. We are more interested in technology; in the how rather than the why. We send our children to advanced schools of education where they will not be required to think about the nature of life or the world. They are only required to think, *"How do you build a better computer?" "How do you make more money?" "How do you design a more obsolete car?"*

The idea of sitting for 30 years and contemplating the nature of life is not very appealing in our time. Yet, for thousands of years in the Western world that was the ultimate job. A philosopher held an especially high place in the ancient world.

Where did philosophy begin? Jewish tradition says it began with **King Solomon**. Many wise men from Athens came to him to test his wisdom, and it was he who got them started on these ideas.

We can perhaps understand this better by studying Solomon's, Ecclesiastes, which is the first book of philosophy. It takes all the other philosophies at the time – e.g. Hedonism, Fatalism and even what later on would be called Epicureanism – and draws them out to their ultimate illogical conclusion. Solomon examines all the possible philosophical answers that exist in the world and does away with each of them: why this does not work and why that does not work. In effect, he shows you where the blanket is too short, even in the best philosophies.

Solomon plays the *devil's advocate* for every philosophical theory. He describes it, and sometimes even seems to indulge in it, but eventually pulls back and points out its fatal flaw. After all is heard, he concludes, the attempts to arrive at a unified philosophy to explain all of life logically is vain and empty.

By extension that naturally leads to the necessity of faith and belief in an Infinite Being whose ways are ultimately beyond the grasp of mere mortals possessed of finite minds.

In the Jewish viewpoint, philosophy is really just an adjunct of Torah. Even though it started its course in Western civilization under Jewish auspices, the Jewish people never really developed it. It was given away to *"the wise men of Athens."* The Greeks absorbed Solomon's methodology and spirit of inquiry, took it back with them to Greece and developed it in their own ways.

In ancient Greece, the philosopher was the most respected member of the community - and the most dangerous. That is why Socrates was put to death. Today, there would be no reason to put Socrates to death. He would not be much of a threat. However, in the Greek world, a philosopher was held in such esteem that he was the single most subversive member of society if he wished to be. He alone could undermine the government.

The Burden of Conscience and the Greeks

The Greeks also developed the most advanced system of paganism in the world. Anyone who has ever read Greek mythology understands what happened. They took all of the bad habits of human beings and gave them to the gods. Instead of humans behaving like God, as the Torah demands (Deuteronomy 28:9), they had the gods behaving like humans.

There was a method to that madness, because it absolved people of responsibility for their sins. If the gods themselves lied, cheated, committed adultery and stole – if they did everything ungodly – they why should mere mortals be expected to act better? Therefore, the ancient Greeks never had the burden of conscience — which not only did the Jewish people have, but even the early pagans had.

"Eat, drink, and be merry, for tomorrow we die," is attributed to the Epicureans, a well-known group of Greek

philosophers. It is an oversimplification, but that represented the Greek idea.

We may think that our society is the same; however, it still has the conscience of the Puritans hanging over it, and, therefore, the more we eat, drink, and be merry the more frightened we are that tomorrow we die.

That dissonance goes a long way in explaining a great deal of the reason for our society's mental illness problems.

Even though we behave in an unprincipled fashion, deep down there is enough of a conscience left over from Christianity and the Western world's system of morality that prevents us from really enjoying ourselves.

The Greeks had none of those hang-ups. And that is one reason why it was a very appealing philosophy and way of life.

Greek Beauty

The Greeks exalted the human body, and in so doing removed all barriers to nudity and sexual behavior. They made aesthetics not just an art form but a form of worship.

Anyone who has seen the ruins of the Parthenon gets a glimpse how magnificent Greek architecture was. Indeed, the Greeks developed mathematics because in their engineering for improved architecture they figured out new forms. They made the temples and the buildings of all other cultures look puny and ugly in comparison.

The Greeks developed music and of dance. They developed musical instruments and played their music with  abandon, often to the accompaniment of ecstatic and/or orgiastic dance.

The Greeks wrote epic poems and plays. The epic poems of Homer, The Iliad and The Odyssey, even today form the basis of Western civilization's literature. The ability to tell a story, to make it rhyme, had a tremendous appeal. Especially to the Jewish people, who were intelligent, literate, super-critical, and constantly looking for something. The nature of the Jew is always to be dissatisfied, to have a soul on fire burning for something more than the world has to offer; a soul ever-thirsting for fulfillment and transcendence.

For all these reasons, when the Greeks burst on the horizon the Jews were very receptive. And in that lay the great danger for the complete destruction of the edifice of the Judaism, the basic ideas and value systems, the observance of the commandments, the study of Torah, etc.

While many Jews would succumb to the influence others would withstand the test. They would come to see Greek enlightenment as darkness, and to realize that what seemed beautiful on the outside was really only skin deep. Beneath the surface was just another frightening predatory beast (Daniel 7:6).

The Greek exile began on a high note. That made it even more of a test, more of a temptation. After the initial euphoria, the Jews found themselves in a clash of cultures; in a world of outer and inner conflict, sometimes with Greeks and other non-Jewish elements, and sometimes with fellow Jews.

The Greek exile would prove different than every other exile Jewish exile. Indeed, it would be an exile while the Jews still lived in their own land. It would be as much an exile of mind as of the body. It would test Jewish resolve like never before, and force the Jewish people to choose between complete disintegration and digging deep into themselves to find an inner enlightenment no Greek poetry or beauty could ever possibly to tap.

If the statistics are right, the Jews constitute but one percent of the human race. It suggests a nebulous dim puff of star dust lost in the blaze of the Milky Way. Properly the Jew ought hardly to be heard of, but he is heard of, has always been heard of. He is as prominent on the planet as any other people, and his commercial importance is extravagantly out of proportion to the smallness of his bulk. His contributions to the world's list of great names in literature, science, art, music, finance, medicine, and abstruse learning are also away out of proportion to the weakness of his numbers. He has made a marvelous fight in the world, in all the ages; and has done it with his hands tied behind him. He could be vain of himself, and be excused for it.

The Egyptian, the Babylonian, and the Persian rose, filled the planet with sound and splendor, then faded to dream-stuff and passed away; the Greek and the Roman followed, and made a vast noise, and they are gone; other peoples have sprung up and held their torch high for a time, but it burned out, and they sit in twilight now, or have vanished.

The Jew saw them all, beat them all, and is now what he always was, exhibiting no decadence, no infirmities of age, no weakening of his parts, no slowing of his energies, and no dulling of his alert and aggressive mind. All things are mortal but the Jew; all other forces pass, but he remains.

In Toras Ha'Olah (1:11:4 – authored by Rabbi Moses Isserles, the "*Ramah*" (the great 16th century leader of world Jewry) it is written: "*For in truth, all the wisdom of the philosophers and researchers came from Israel, and all of their wisdom is encompassed in the Torah, as the Rabbi of the "Guide for the Perplexed," i.e. Maimonides) taught at length (1:71)* .... The wisdom of Aristotle was stolen from King Solomon, for when Alexander the Macedonian conquered Jerusalem he set his teacher Aristotle to govern over the collection of the books of Solomon.

"*And every good thing he found in them he copied and intermixed in them some his own mistaken ideas, such as the Antiquity of the World and the denial of Providence....*"

Chapter Seven

The Greek Bible

The Septuagint (pronounced 'sep choo a jint) is sometimes known as LXX or the Greek Old Testament, is an ancient translation of the Hebrew Bible and some related texts into Koine Greek (Alexandrian Greek), dated as early as the late 2nd century BCE. It is quoted in the New Testament, particularly in the writings of Paul the Apostle, and also by the Apostolic Fathers and later Greek Church Fathers, and continues to serve as the Eastern Orthodox Old Testament.

The LXX in Bible commentary refers to the Septuagint Commentary which is the oldest Greek version of the Old Testament. It is said to have been translated from the Hebrew by Jewish scholars in 3 or 2 BCE. It was translated to meet the needs of Greek-speaking Jews who lived outside of Palestine. Some say that it has to do with the number seventy because it is thought that there were seventy translators who supposedly worked for 70 days. In Roman Numbers L=50 and X=10 so LXX=70.

The traditional story is that Ptolemy II sponsored the translation for use by the many Alexandrian Jews who were not fluent in Hebrew but fluent in Koine Greek, which was the main language in Alexandria, Egypt and the   Eastern Mediterranean from the death of Alexander the Great in 323 BCE until the development of Byzantine Greek around 600 CE.

The Septuagint should not be confused with the seven or more other Greek versions of the Old Testament, most of which did not survive except as fragments (some parts of these being known from Origen's Hexapla, a comparison of six translations in adjacent columns, now almost wholly lost). Of these, the most important are the three: those by Aquila, Symmachus, and Theodotion.

Earlier Greek language

The date of the 3rd century BCE, given in the traditional story is confirmed (*for the Torah translation*) by a number of factors, including the Greek being representative of early Koine, citations beginning as early as the 2nd century BCE, and early manuscripts datable to the 2nd century. That from today would be about 2,214 years ago.

After the Torah, other books were translated over the next two to three centuries. It is not altogether clear which was translated when, or where; some may even have been translated twice, into different versions, and then revised. The quality and style of the different translators also varied considerably from book to book, from the literal to paraphrasing to interpretative.

The translation process of the Septuagint can be broken down into several distinct stages, during which the social milieu of the translators shifted from Hellenistic Judaism to Early Christianity. The translation began in the 3rd century BCE and was completed by 132 BCE, initially in Alexandria, but in time elsewhere as well. The Septuagint is the basis for the Old Latin, Slavonic, Syriac, Old Armenian, and Old Georgian and Coptic versions of the Christian Old Testament.

Some sections of the Septuagint may show Semitisms, or idioms and phrases based on Semitic languages like Hebrew and Aramaic. Other books, such as the Daniel and Proverbs, show Greek influence more strongly. Jewish Koine Greek exists primarily as a category of literature, or cultural category, but apart from some distinctive religious vocabulary is not distinct from other varieties of Koine Greek to be counted a separate dialect of Greek.

The Septuagint is also useful for elucidating pre-Masoretic Hebrew: many proper nouns are spelled out with Greek vowels in the LXX, while contemporary Hebrew texts lacked vowel pointing. One must, however, evaluate such - evidence with caution since it is extremely unlikely that all ancient Hebrew sounds had precise Greek equivalents.

Disputes over Canonicity

Historically there were some disputes over canonicity. As the work of translation progressed, the canon of the Greek Bible expanded. The Torah (*Pentateuch in Greek*) always maintained its pre-eminence as the basis of the canon; but the collection of prophetic writings, based on the

Jewish Nevi'im, had various hagiographical works incorporated into it.

Newer books added

In addition, some newer books were included in the Septuagint: those called anagignoskomena (to be read) in Greek, because they are not included in the Jewish canon. Among these are the Maccabees and the Wisdom of Ben Sira. Also, the Septuagint version of some Biblical books, like Daniel and Esther, are longer than those in the Jewish canon. Some of these apocryphal (*questionable authenticity) books (e.g. the Wisdom of Solomon, and the second book of Maccabees*) were not translated, but composed directly in Greek.

Septuagint missing Jewish Documents

The canonicity of the larger group of writings (*the Jewish ketuvim*) had not yet been established although some sort of selective process must have been employed because the Septuagint did not include other well-known Jewish documents such as Enoch or Jubilees or other writings that are not part of the Jewish canon. These are now classified as Pseudepigrapha since Late Antiquity, once attributed to a Council of Jamnia, mainstream rabbinic Judaism rejected the Septuagint as valid Jewish scriptural texts.

Several reasons have been given for this. First, some mistranslations were claimed Septuagint and is evident in the Old Testament of earliest Christian Bibles (4th century). Some books that are set apart in the Masoretic text are grouped together. For example, the Books of Samuel and the Books of Kings are in the LXX one book in four parts called Βασιλειῶν (*Of Reigns*). In LXX, the Books of Chronicles Supplement Reigns and it is called Paraleipoménon (Παραλειπομένων— (*things left out*). The Septuagint organizes the Minor Prophets as twelve parts of one Book of Twelve.

Fabricated Scriptures

Some scripture of ancient origin is found in the Septuagint but are not present in the Hebrew. These additional books are Tobit, Judith, Wisdom of Solomon, Wisdom of Jesus son of Sirach, Baruch, Letter of Jeremiah (which later became chapter 6 of Baruch in the Vulgate), additions to Daniel, The Prayer of Azarias, the Song of the Three Children, Susanna and Bel and the Dragon), additions to Esther, 1 Maccabees, 2 Maccabees, 3 Maccabees, 4 Maccabees, 1 Esdras, Odes, including the Prayer of Manasseh, the Psalms of Solomon, and Psalm 151.

Varied Acceptance

The canonical acceptance of these books varies among different Christian traditions, and there are canonical books not derived from the Septuagint.

In most ancient copies of the Bible which contain the

Septuagint version of the Old Testament, the Book of **Daniel** is Theodotion's translation from the Hebrew, which more closely resembles the Masoretic text. The Septuagint version was discarded in favor of Theodotion's version in the 2nd to 3rd centuries CE. In Greek-speaking areas, this happened near the end of the 2nd century, and in Latin-speaking areas (*at least in North Africa*), it occurred in the middle of the 3rd century. History does not record the reason for this, and St. Jerome reports, in the preface to the Vulgate version of Daniel, *"This thing 'just' happened."*

Greek version of Daniel

One of two Old Greek texts of the **Book of Daniel** has been recently rediscovered and work is ongoing in reconstructing the original form of the book. Not the original Septuagint version, but instead is a copy of the canonical Ezra-Nehemiah is known in the Septuagint as *"Esdras B"*, and 1 Esdras is *"Esdras A"*. 1 Esdras is a very similar text to the books of Ezra-Nehemiah, and the two are widely thought by scholars to be derived from the same original text. It has been proposed, and is thought highly likely by scholars, that "*Esdras B*" – the canonical Ezra-Nehemiah – is Theodotion's version of this material, and "*Esdras A*" is the version which was previously in the Septuagint on its own. The Early Christian Church used the Greek texts since Greek was a main language of the Roman Empire at the time, and the language of the Greco-Roman Church (Aramaic was the language of Syrian Christianity, which used the Targums (interpretations).

Early Christian Church uses Greek version

Many Hebrew Texts lost

The relationship between the apostolic use of the Old Testament, for example, the Septuagint and the now lost

Hebrew texts (though to some degree and in some form carried on in Masoretic tradition) is complicated.

The Septuagint seems to have been a major source for the Apostles, but it is not the only one. St. Jerome offered, for example, Matt 2:15 and 2:23, John 19:37, John 7:38, 1 Cor. 2:9 as examples not found in the Septuagint, but in Hebrew texts. (Matt 2:23 is not present in current Masoretic tradition either, though according to St. Jerome it was in Isaiah 11:1.)

New Testament based on Greek

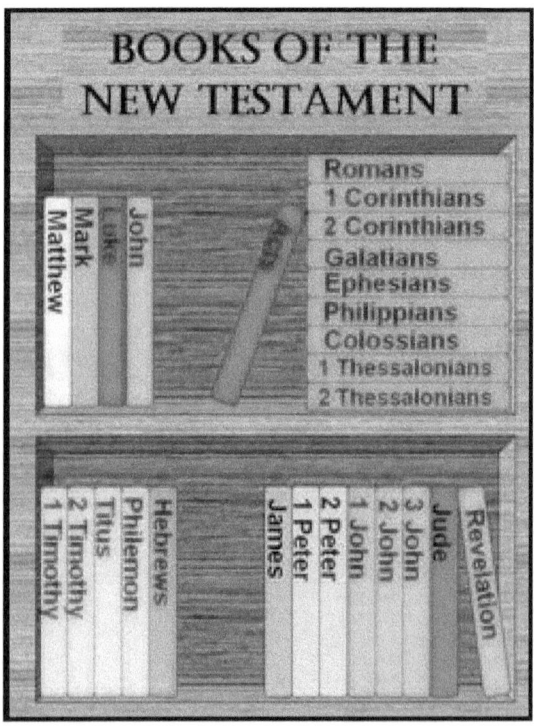

The New Testament writers, when citing the Jewish scriptures, or when quoting Jesus doing so, freely used the Greek translation, implying that Jesus, his Apostles and their followers considered it reliable.

100

In the Early Christian Church, the presumption that the Septuagint was translated by Jews before the era of Christ, and that the Septuagint at certain places gives itself more to an Christological interpretation than 2nd-century Hebrew texts was taken as evidence that Jews had changed the Hebrew text in a way that made them less Christological.

For example, **Irenaeus** concerning Isaiah 7:14: The Septuagint clearly writes of a *virgin* that shall conceive. while the Hebrew text was, according to Irenaeus, at that time interpreted by Theodotion and Aquila (both proselytes of the Jewish faith) as a *young woman* that shall conceive.

According to Irenaeus, the Ebionites used this to claim that Joseph was the (*biological*) father of Jesus. From Irenaeus' point of view that was pure heresy, facilitated by (late) anti-Christian alterations of the scripture in Hebrew, as evident by the older, pre-Christian, Septuagint.

Pre-Christian Septuagint

When Jerome undertook the revision of the Old Latin translations of the Septuagint, he checked the Septuagint against the Hebrew texts that were then available. He broke with church tradition and translated most of the Old Testament of his Vulgate from Hebrew rather than Greek.

His choice was severely criticized by **Augustine**, his contemporary; the story of a flood was taken as evidence that Jews had changed the Hebrew text in a way that made them less Christological than 2nd-century Hebrew texts and still less moderate criticism came from those who regarded Jerome as a forger. While on the one hand he argued for the superiority of the Hebrew texts in correcting the Septuagint on both philological and theological grounds, on the other, in the context of accusations of heresy against him, Jerome would acknowledge the Septuagint texts as well. With the passage of time, acceptance of Jerome's version gradually increased until it displaced the Old Latin translations of the Septuagint.

The Eastern Orthodox Church still prefers to use the LXX as the basis for translating the Old Testament into other languages. The Eastern Orthodox also use LXX untranslated where Greek is the liturgical language, in the Orthodox Church of Constantinople, the Church of Greece and the Cypriot Orthodox Church.

Critical translations of the Old Testament, *critics while using the Masoretic Text as their basis, consult the* ***Septuagint*** *as well as other versions in an attempt to reconstruct the meaning of the Hebrew text whenever the latter is unclear, undeniably corrupt, or ambiguous.*

The Septuagint (Greek Bible) was based on the Hebrew Scriptures and became the first Christian Bible.

Chapter Eight

A brief History of the Many Bibles

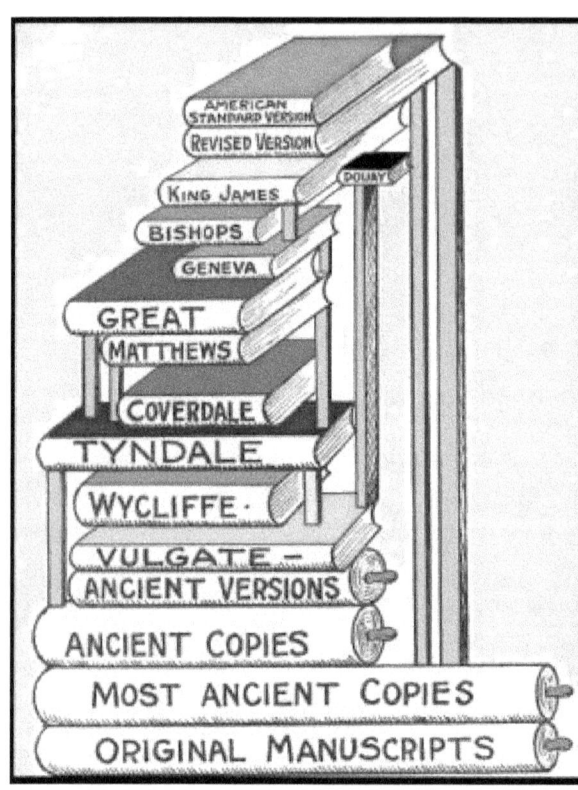

The **first Bible** in history was the **Tanakh** (*pronounced ta-nock*) which is the name used in Judaism for the canon of the Hebrew Scriptures.

The **Septuagint** (pronounced 'sep CHo͞o a jint) is sometimes known as LXX or Greek Old Testament, is an ancient translation of the Hebrew Bible and some related texts into Koine Greek, dated as early as the late 2nd century BCE. (*About 2,214 years ago*)

The **Vulgate,** also known as *"The Latin Bible"* For over a thousand years (400 - 1530 CE), the Vulgate was the definitive edition of the most influential text in Western European society.

The **Luther Bible** is a German language Bible translation from Hebrew and ancient Greek by Martin Luther, of which the New Testament was published in 1522 and the complete Bible, containing the Old and New Testaments and Apocrypha, in 1534.

The **Tyndale Bible** generally refers to the body of biblical translations by William Tyndale. Tyndale's Bible is credited with being the first English translation to work directly from Hebrew and Greek texts.

The **Great Bible** was the first authorized edition of the Bible in English, authorized by King Henry VIII of England to be read aloud in the church services of the Church of England. The Great Bible was prepared by Myles Coverdale, working under commission of Thomas, Lord Cromwell, and Secretary to Henry VIII and Vicar General.

The **Bishops' Bible** is an English translation of the Bible which was produced under the authority of the established Church of England in 1568. It was substantially revised in 1572, and this revised edition was to be prescribed as the base text for the Authorized King James Version of 1611.

"*Breeches Bible*" is a book-collectors' term for the **Geneva Bible** of 1560. The term derives from the reference in Genesis iii: 7 to Adam and Eve clothing themselves in "*breeches*" made from fig leaves. The New Testament appeared in 1557, and was probably the product of one man, William Whittingham, an Englishman of great learning, and related to Calvin by marriage. It was a revision of Tyndale's (*1525*), with an introduction by Calvin. It was the first to use the division of the text into verses.

The **King James Authorized Version** was meant to replace the Bishops' Bible as the official version for readings in the Church of England.

No record of its authorization exists; it was probably affected by an order of the Privy Council but the records for the years 1600 to 1613 were destroyed by fire in January 1618/19 and it is commonly known as the Authorized Version in the United Kingdom.

The King's Printer issued no further editions of the Bishops' Bible, so necessarily the Authorized Version replaced it as the standard lectern Bible in parish church use in England.

Many people are familiar with the Bible, but few people are aware of the countless versions and revisions it has undergone since it was first written.

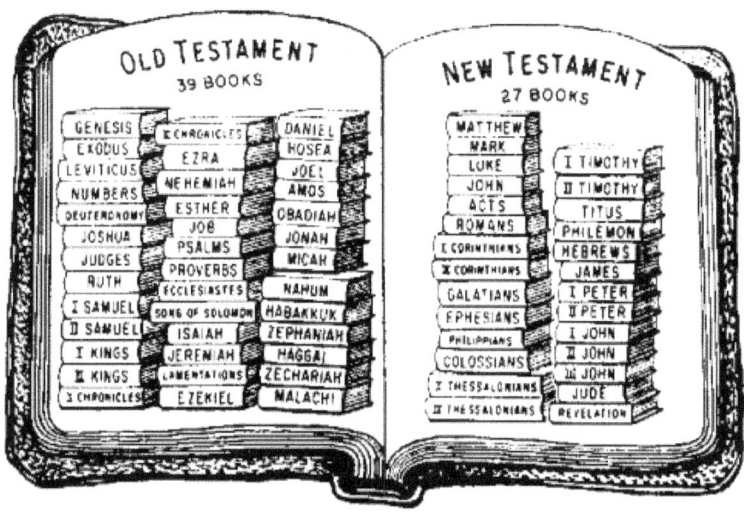

The first part of Christian Bibles is the Old Testament, which contains, at minimum, the twenty-four books of the Hebrew Bible divided into thirty-nine books and ordered differently than the Hebrew version Bible. The

Catholic Church and Eastern Christian churches versions also hold certain deuterocanonical books and passages to be part of the Old Testament canon.

The second part is the New Testament version, containing twenty-seven books; the four Canonical gospels, the Acts of the Apostles, twenty-one Epistles or letters and the Book of Revelation.

By the 2nd century BCE Jewish groups had called the books of the Bible *"holy,"* and Christians now commonly call the Old and New Testaments of the  Christian Bible *"The Holy Bible"* or ***"the Holy Scriptures"***. Many Christians consider the whole canonical text of the Bible to be divinely inspired.

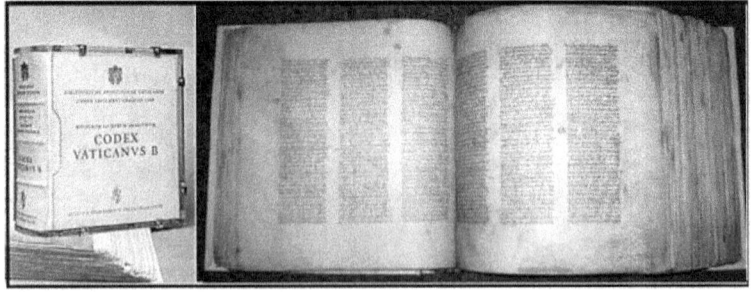

The oldest surviving complete Christian Bibles are Greek manuscripts from the 4th century, about 2,414 years ago. An early 4th-century Septuagint translation is found in the **Codex Vaticanus**. It is so called because it belongs to the Vatican Library (*Codex Vaticanus, 1209*).

The Greek manuscript is considered the most important of all the manuscripts of Holy Scripture. This codex is an approximately eight section volume written in all lower-case letters of the fourth century, on folios of fine parchment bound together five to a group. Each page is divided into three columns of forty lines each, with from sixteen to eighteen letters to a line, except in the poetical books, where, owing to the fixed division of the lines, there are but two columns to a page. There are no capital letters, but at times the first letter of a section extends over the margin. Several hands worked at the manuscript; the first writer inserted neither pauses nor accents, and made use but rarely of a simple punctuation.

Unfortunately, the codex is mutilated; at a later date the missing folios were replaced by others. Thus, the first twenty original folios are missing; a part of folio 178, and ten folios after folio 348; whose exact number it is impossible to establish.

There are in all, 759 original folios. The contents of the New Testament deal explicitly with first century Christianity. Therefore, the New Testament has frequently accompanied the spread of Christianity around the world.

It reflects and serves as a source for Christian theology. Both extended readings and phrases directly from the New Testament, along with readings from the Old Testament, are also incorporated into the various Christian liturgies.

The New Testament has influenced not only religious, philosophical, and political movements in Christendom, but also has left an influence on literature, art, and music.

The New Testament is an anthology, a collection of Christian works written in the common Greek language of the first century at different times by various writers, who were early Jewish disciples of Jesus of Nazareth. In almost all Christian traditions today, the New Testament consists of 27 books. The original texts were written in the first and perhaps the second centuries of the Christian Era, generally believed to be in Koine Greek, which was the common language of the Eastern Mediterranean from the Conquests of Alexander the Great *(335 - 323 BCE)* until the evolution of Byzantine Greek. All of the works which would eventually be incorporated into the New Testament would seem to have been written no later than around 150 CE. *(About 1,864 years ago)*

Collections of related texts such as letters of the Apostle Paul (*a major collection of which must have been made already by the early 2nd century*) and the Canonical Gospels of Matthew, Mark, Luke, and John asserted by Irenaeus of Lyon in the late-2nd century as the Four Gospels gradually were joined to other collections and single works in different combinations to form various Christian canons of Scripture.

Over time, some disputed books, such as the Book of Revelation and the Minor Catholic (General) Epistles were introduced into canons in which they were originally absent. Other works earlier held to be Scripture, such as 1

Clement, the Shepherd of Hermas, and the Diatessaron, were excluded from the New Testament.

The Old Testament canon is not completely uniform among all major Christian groups including Roman Catholics, Protestants, the Greek Orthodox Church, the Slavic Orthodox Churches, and the Armenian Orthodox Church.

However, the twenty-seven-book canon of the New Testament, at least since Late Antiquity, has been almost universally recognized within Christianity.

The Gospels: include four narratives of the life teaching and death of Jesus, composed by Mathew, Mark, Luke and John.

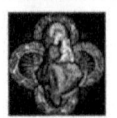

The *"Acts of the Apostles,"* a narrative of the Apostles' in the early church, probably written by the same writer as the Gospel of Luke, which it continues. The *"Epistles,"* are twenty-one letters written by various authors and consists of Christian doctrine, counsel, instruction, and conflict resolution. And the *"Book of Revelation,"* an Apocalypse, which is a book of prophecy, containing some instructions to the seven local congregations of Asia Minor, but mostly containing prophetical symbiology, about the end times

The divide between Jews and Christians has been built by thousands of years of controversy.

For example, in 336 AD the 30-member Catholic Council of Laodicea decided that Sunday would be a better day for the **Sabbath** and changed it. Thus, from that day forward Christians celebrated the Sabbath on Sunday while the Jews continued with celebrating the Sabbath as Saturday. There have been volumes of evidence compiled illustrating how this single act not only defies the accepted calendar days of the week but biblical references throughout the Bible itself and even many laws in Israel.

Obviously, it remains a big point of contention between Christians and Jews. For instance, in Israel you can do many things, commit many sins and still remain a Jew. However, a **Jew who converts to Christianity** is guilty of such a tremendous crime that members of the perpetrator's own family will put his name on a casket and hold a funeral for him/her. From that day forward they are ostracized from the church and many times the family as well.

Throughout history the Bible has been translated, and rewritten many times, and each new translation of the Bible has been reviewed by the leaders at the time. Once an "*acceptable*" version was established it seems it was translated into most every major language in the world.

The bible's influence has always been staggering by its most uniformly acceptance by the people of most nations. Historically the most complicated controversial translations have been the English versions. In the past translators have been locked in towers with the original work such as manuscripts and scrolls and a supply of pens and ink – and not been let out until a suitable translation pleased the leaders.

Others have completed their work and then been sent to prison and one (Tyndale) was even choked to death and burned at the stake.

The Christian Bible was divided into chapters in the 13th century by Stephen Langton and into verses in the 16th century by French printer Robert Estienne and is now usually cited by book, chapter, and verse.

To be sure, religion has the unparalleled power to bring people into groups. Religion has helped humans survive, adapt and evolve in groups over the ages. It's also helped us learn to cope with death, identify danger and even finding mating partners.

The Gutenberg Bible was the first printed Bible, since then the Bible has estimated annual sales of 25 million copies, and has been a major influence on literature and history, especially in the West where it was the first mass printed book.

There are those who say organized religion is a scam that takes advantage of an already existing human trait to desire things to believe in. Yet there are others who say, *"we are giving people something good and wonderful to focus their beliefs on, spreading happiness, fellowship and even the power to heal both the human body and the spirit!"*

It seems all a matter of perspective

The truth of the matter is very simple, and that is when people like Christians, Jews or Muslims all claim to be descendants of Abraham, they are all probably right. And no matter by what name you call God, or in what manner you worship, you are fulfilling your inherited instinct to search for and find something to believe in. The bottom line is you were born this way, as were your ancestors and as will be your descendants. It is purely human nature in its most basic form. It is what it is.

Chapter Nine

What other religions say about faith

Baha'i Faith:

In the Baha'i Faith, faith is ultimately the acceptance of the divine authority of the Manifestations of God. In the religion's view, faith and knowledge are both required for spiritual growth. Faith involves more than outward obedience to this authority, but also must be based on a deep personal understanding of religious teachings. By faith is meant, first, conscious knowledge, and second, the practice of good deeds.

Buddhism:

Faith (Pali: Saddhā, Sanskrit: Śraddhā) is an important constituent element of the teachings of Gautama Buddha in both the Theravada and the Mahayana traditions. The teachings of Buddha were originally recorded in the language Pali and the word saddhā is generally translated as "*faith*". In the teachings, saddhā is often described as:

- A conviction that something is.

- A determination to accomplish one's goals.

- A sense of joy deriving from the other two.

While faith in Buddhism does not imply "*blind faith*", Buddhist practice nevertheless requires a degree of trust, primarily in the spiritual attainment of Gautama Buddha. Faith in Buddhism centers on the understanding that the Buddha is an awakened being, on his superior role as a teacher, in the truth of his Dharma (spiritual teachings) and in his Sangha (community of spiritually developed followers).

Faith in Buddhism can be summarized as faith in the **Three Jewels:**

- Buddha
- Dharma
- Sangha.

It is intended to lead to the goal of enlightenment, orbodhi, and Nirvana.

Volitionally, faith implies a resolute and courageous act of will. It combines the steadfast resolution that one will do a thing with the self-confidence that one can do it.

As a counter to any form of "*blind faith*", the Buddha's teachings included those included in the Kalama Sutra, exhorting his disciples to investigate any teaching and to live by what is learnt and accepted, rather than believing in something simply because it is taught.

Christianity

Faith in Christianity is based on the work and teachings of **Jesus Christ.** Christianity declares not to be distinguished by faith, but by the object of its faith. Rather than being passive, faith leads to an active life aligned with the ideals and the example of the life of Jesus. It sees the mystery of God and his grace and seeks to know and become obedient to God. To a Christian, faith is not static but causes one to learn more of God and grow, and has its origin in God.

In **Christianity**, faith causes change as it seeks a greater understanding of God. Faith is not fideism or simple obedience to a set of rules or statements. Before Christians have faith, they must understand in whom and in what they have faith.

Without understanding, there cannot be true faith, and that understanding is built on the foundation of the community of believers, the scriptures and traditions and on the personal experiences of the believer. In English translations of the New Testament, the word faith generally corresponds to the Greek noun πίστις (pistis) or the Greek verb πιστεύω (pisteuo) meaning: *"to trust, to have confidence, faithfulness, to be reliable, to assure".* The Christian Bible says that faith is *"the substance of things hoped for, the evidence of things not seen."*

Hinduism

All schools of Hindu philosophy posit that consciousness (ātman) is distinct and independent from mind and matter (prakṛti). Therefore, Hindu faith is based on the premise that logic and reason are not conclusive methods of epistemic knowing. Spiritual practice (sadhana) is performed with the faith that knowledge beyond the mind and sense perception will be revealed to the practitioner. Śraddhā (ITRANS: shraddhA) is translated as faith in Sanskrit.

The schools of Hindu philosophy differ in their recommended methods to cultivate faith, including selfless action (karma-yoga), renunciation (jnana-yoga) and devotion (bhakti-yoga).

In chapter 17 of the **Bhagavad Gita**, Krishna describes how faith, influenced by the three modes (guṇas) lead to different approaches in worship, diet, sacrifice, austerity and charity. Swami Tripurari states: Faith for good reason arises out of the mystery that underlies - the very structure and nature of reality, a mystery that in its entirety will never be entirely demystified despite what those who have placed reason on their altar might like us to believe. The mystery of life that gives rise to faith as a supra-rational means of unlocking life's mystery - one that reason does not hold the key to - suggests that faith is fundamentally rational in that it is a logical response to the mysterious.

Islam

Faith is complete submission to the will of God, which includes belief, profession and the body's performance of deeds, consistent with the commission as vicegerent on Earth; all according to God's will towards man has two aspects:

1.) Recognizing and affirming that there is one Creator of the universe and only to this Creator is worship due. According to Islamic thought, this comes naturally because faith is an instinct of the human soul. This instinct is then trained via parents or guardians into specific religious or spiritual paths. Likewise, the instinct may not be guided at all.

2.) Willingness and commitment to submitting that God exists, and to His prescriptions for living in accordance with vicegerency. The Qur'an is understood as the dictation of God's prescriptions through the Prophet Muhammad and is believed to have been updated and completed the previous revelations that God sent through earlier prophets.

In the **Qur'an,** it is stated that (2:62): "Surely, those who believe, those who are Muslims, Jewish, the Christians, and the Sabians; anyone who (1) believes in GOD, and (2) believes in the Last Day, and (3) leads a righteous life, will receive their recompense from their Lord. They have nothing to fear, nor will they grieve."

Judaism

In **Jewish principles of faith**, faith itself is not a religious concept in Judaism. Although Judaism does recognize the positive value of Emunah (generally translated as faith, trust in God) and the negative status of the Apikorus (heretic), faith is not as stressed or as central as it is in other religions, especially compared with Christianity and Islam. It could be a necessary means for being a practicing religious Jew, but the emphasis is placed on practice rather than on faith itself. Very rarely does it relate to any teaching that must be believed. Classical Judaism does not require one to explicitly identify God (a key tenet of faith in Christianity), but rather to honor the idea of God.

In the Jewish scriptures trust in God - Emunah - refers to how God acts toward his people and how they are to respond to him; it is rooted in the everlasting covenant established in the Torah, notably Deuteronomy 7:9 *"Know, therefore, that only the LORD your God is God, the steadfast God who keeps his gracious covenant to the thousandth generation of those who love Him and keep His commandments"*

The specific tenets that compose required belief and their application to the times have been disputed throughout Jewish history. Today many, but not all, Orthodox Jews have accepted Maimonides' Thirteen Principles of Belief.

A traditional example of Emunah as seen in the Jewish annals is found in the person of Abraham. On a number of occasions, Abraham both accepts statements from God that seem impossible and offers obedient actions in response to direction from God to do things that seem implausible (see Genesis 12-15).

"The Talmud describes how a thief also believes in God: On the brink of his forced entry, as he is about to risk his life—and the life of his victim—he cries out with all sincerity, 'God help me!' The thief has faith that there is a God who hears his cries, yet it escapes him that this God may be able to provide for him without requiring that he abrogate God's will by stealing from others. For emunah to affect him in this way he needs study and contemplation."

Sikh

Sikhism, the fifth-largest organized religion in the world, was founded in 15th-century Punjab on the teachings of Guru Nanak Dev and ten successive Sikh gurus, the last one being the sacred text Guru Granth Sahib. The core philosophy of the Sikh religion is described in the beginning hymn of the Guru granth Sahib, *"There is one supreme eternal reality; the truth; imminent in all things; creator of all things; immanent in creation. Without fear and without hatred; not subject to time; beyond birth and death; self-revealing."*

Guru Nanak, the founder of the faith, summed up the basis of Sikh lifestyle in three requirements: 1.) Nām Japō (meditate on the holy name 2.) (Waheguru), Kirat karō (work diligently and honestly) and 3.) Vaṇḍ chakkō (share one's fruits).

127

Meher Baba

Meher Baba described three types of faith, emphasizing the importance of faith in a spiritual master:

1.) Faith in oneself

2.) Faith in the Master

3.) Faith in life.

"Faith is so indispensable to life that unless it is present in some degree, life itself would be impossible. It is because of faith that cooperative and social life becomes possible. It is faith in each other that facilitates a free give and take of love, a free sharing of work and its results.

When life is burdened with unjustified fear of one another it becomes cramped and restricted.... Faith in the Master becomes all-important because it nourishes and sustains faith in oneself and faith in life in the very teeth of set-backs and failures, handicaps and difficulties, limitations and failings. Life, as man knows it in himself, or in most of his fellow-men, may be narrow, twisted and perverse, but life as he sees it in the Master is unlimited, pure and untainted.

In the Master, man sees his own ideal realized; the Master is what his own deeper self would rather be. He sees in the Master the reflection of the best in himself which is yet to be, but which he will surely one day attain. Faith in the Master therefore becomes the chief motive-power for realizing the divinity which is latent in man."

Epistemological validity of faith

Epistemology is the branch of philosophy that studies knowledge. It attempts to answer the basic question: what distinguishes true (adequate) knowledge from false (inadequate) knowledge? Practically, this question translates into issues of scientific methodology: how can one develop theories or models that are better than competing theories? It also forms one of the pillars of the new sciences of cognition, which developed from the information processing approach to psychology, and from artificial intelligence, as an attempt to develop computer programs that mimic a human's capacity to use knowledge in an intelligent way.

There is a wide spectrum of opinion with respect to the epistemological validity of faith. On one extreme is logical positivism, which denies the validity of any beliefs held by faith; on the other extreme is fideism, which holds that true belief can only arise from faith, because reason and physical evidence cannot lead to truth.

Some foundationalists, such as St. Augustine of Hippo and Alvin Plantinga, hold that all of our beliefs rest ultimately on beliefs accepted by faith. Others, such as C.S. Lewis, hold that faith is merely the virtue by which we hold to our reasoned ideas, despite moods to the contrary.

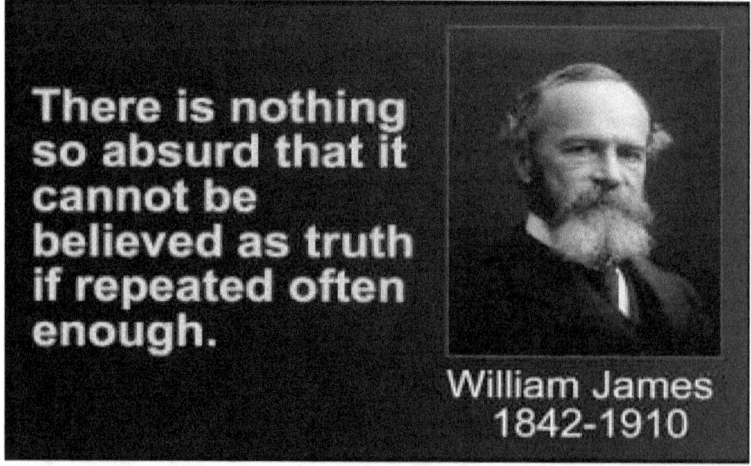

There is nothing so absurd that it cannot be believed as truth if repeated often enough.

William James
1842-1910

William James believed that the varieties of religious experiences should be sought by psychologists, because they represent the closest thing to a microscope of the mind - that is, they show us in drastically enlarged form the normal processes of things. For a useful interpretation of human reality, to share faith experience he said that we must each make certain "*over-beliefs*" in things which, while they cannot be proven on the basis of experience, help us to live fuller and better lives.

Fideism

Fideism: is an epistemological theory which maintains that faith is independent of reason, or that reason and faith are hostile to each other and faith is superior at arriving at particular truths (see natural theology). The word fideism comes from fides, the Latin word for faith, and literally means "*faith-ism.*"

Fideism is not a synonym for religious belief, but describes a particular philosophical proposition in regard to the relationship between faith's appropriate jurisdictions at arriving at truths, contrasted against reason. It states that faith is needed to determine some philosophical and religious truths, and it questions the ability of reason to arrive at all truth. The word and concept had its origin in the mid- to late-19th century by way of Catholic thought, in a movement called Traditionalism. The Roman Catholic Magisterium has, however, repeatedly condemned fideism.

Support

Religious epistemologists have formulated and defended reasons for the rationality of accepting belief in God without the support of an argument. Some religious epistemologists hold that belief in God is more analogous to belief in a person than belief in a scientific hypothesis. Human relations demand trust and commitment. If belief in God is more like belief in other persons, then the trust that is appropriate to persons will be appropriate to God.

American psychologist and philosopher William James offers a similar argument in his lecture The Will to Believe. Foundationalism is a view about the structure of justification or knowledge. Foundationalism holds that all knowledge and justified belief are ultimately based upon what are called properly basic beliefs. This position is intended to resolve the infinite regress problem in epistemology. According to foundationalism, a belief is epistemically justified only if it is justified by properly basic beliefs. One of the significant developments in foundationalism is the rise of reformed epistemology.

Reformed epistemology is a view about the epistemology of religious belief, which holds that belief in God can be properly basic. Analytic philosophers Alvin Plantinga and Nicholas Wolterstorff develop this view. Plantinga holds that an individual may rationally believe in God even though the individual does not possess sufficient evidence to convince an agnostic – (One who believes that it is impossible to know whether there is a God or one who is skeptical about the existence of God but does not profess true atheism).

One difference between reformed epistemology and fideism is that the former requires defense against known objections, whereas the latter might dismiss such objections as irrelevant. Plantinga has developed reformed epistemology in Warranted Christian Belief as a form of externalism that holds that the justification conferring factors for a belief may include external factors. Some theistic (is the belief that at least one deity exists) philosophers have defended theism by granting evidentialism but supporting theism through deductive arguments whose premises are considered justifiable.

Some of these arguments are probabilistic, either in the sense of having weight but being inconclusive, or in the sense of having a mathematical probability assigned to them. Notable in this regard are the cumulative arguments presented by British philosopher Basil Mitchell and analytic philosopher Richard Swinburne, whose arguments are based on Bayesian probability. In a notable exposition of his arguments, Swinburne appeals to an inference for the best explanation.

Criticism: **Bertrand Russell** noted, *"Where there is*

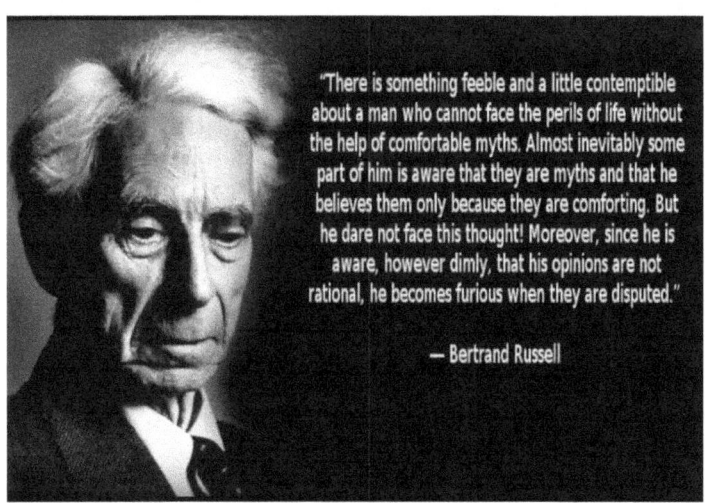

"There is something feeble and a little contemptible about a man who cannot face the perils of life without the help of comfortable myths. Almost inevitably some part of him is aware that they are myths and that he believes them only because they are comforting. But he dare not face this thought! Moreover, since he is aware, however dimly, that his opinions are not rational, he becomes furious when they are disputed."

— Bertrand Russell

evidence, no one speaks of 'faith'. We do not speak of faith that two and two are four or that the earth is round. We only speak of faith when we wish to substitute emotion for evidence. "

Evolutionary biologist **Richard Dawkins** criticizes all faith by generalizing from specific faith in propositions that conflict directly with scientific evidence. He describes faith as mere belief without evidence; a process of active non-thinking. He states that it is a practice that only degrades our understanding of the natural world by allowing anyone to make a claim about nature that is based solely on their personal thoughts, and possibly distorted perceptions, that does not require testing against nature, has no ability to make reliable and consistent predictions, and is not subject to peer review.

Chapter Ten

What Science says about faith

The word faith is often used as a substitute for hope, trust or belief. In religion, faith often involves accepting claims about the character of a deity, nature, or the universe. While some have argued that faith is opposed to reason, proponents of faith argue that the proper domain of faith concerns questions which cannot be settled by evidence. For example, faith can be   applied to predictions of the future, which (*by definition*) has   not yet occurred.

In the movie *"The Greatest Story Ever Told,"* there is an unforgettable scene where Jesus is walking in a hurry to get somewhere and an old man in tattered clothing joins him and grasping his arm he says to Jesus, *"thank you so much for healing me with your powers."* Jesus stops and with his hand on the old man's shoulder he says - *"I have NOT healed you – YOUR faith has healed you."*

Faith is deeply embedded in the human brain, which is programmed for religious experiences, according to a United States study. Religious belief and behavior are a hallmark of human life, with no accepted animal equivalent, and found in all cultures.

Scientists searching for the neural "*Faith switch*," which is used to control religious belief, believe several areas of the brain form the biological foundations of religious belief. The researchers said their findings supported the idea that the brain had evolved to be sensitive to any form of belief that improved the chances of survival, which could explain why a belief in God and the supernatural became so widespread in human evolutionary history.

Research results are unique in demonstrating that specific components of religious belief are mediated by well-known brain networks and they support contemporary psychological theories that ground religious belief within evolutionary-adaptive cognitive functions.

Some evolutionary theorists have suggested that Darwinian natural selection may have put a premium on individuals if they were able to use religious belief to survive hardships that may have overwhelmed those with no religious convictions. Others have suggested that religious belief is a side effect of a wider trait in the human brain to search for coherent beliefs about the outside world. Religion and belief in God, they argue, are just a manifestation of this intrinsic, biological phenomenon that makes the human brain so intelligent and adaptable.

The latest study, involved analyzing the brains of volunteers, who had been asked to think about religious and moral problems and questions. For the analysis, researchers used a *functional MRI machine*, which can identify the most active regions of the brain.

They found that people of different religious persuasions and beliefs, *including atheists*, tended to use the *same electrical circuits* in the brain when solving a moral conundrum as well as when dealing with issues related to God. The study found that several areas of the brain were involved in religious belief, one within the frontal lobes of the cortex - which are unique to humans - and another in the more evolutionary-ancient regions deeper inside the brain, which humans share with apes and other primates.

There is nothing unique about religious belief in these brain structures. Religion doesn't have a 'God spot' as such; instead it's *embedded* in a whole range of other belief systems in the brain that we use every day.

The search for the God spot has in the past led scientists to many different regions of the brain. An early contender was the brain's temporal lobe, a large section that sits over each ear, *because* temporal-lobe epileptics suffering seizures in those regions frequently reported intense religious experiences. One of the principal exponents of this idea asked several of his patients with temporal-lobe epilepsy to listen to a mixture of religious, sexual and neutral words while measuring their levels of arousal and emotional reactions.

Other studies of people taking part in Buddhist meditation suggested the parietal lobes at the upper back region of the brain were involved in controlling religious belief, in particular the mystical elements that gave people a feeling of being on a higher plane during prayer. Religious words elicited an unusually high response in these patients.

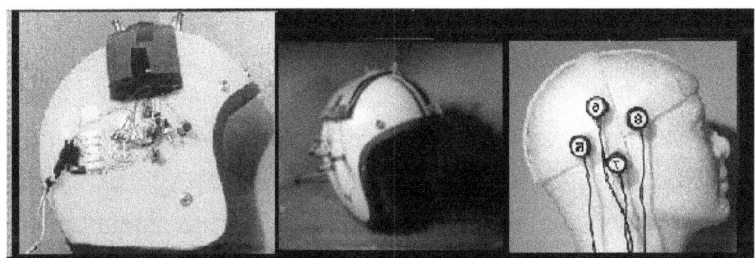

The God Helmet has been used by researchers to study brain activity. This work was followed by a study where scientists tried to stimulate the temporal lobes with a rotating magnetic field produced by a *God helmet* (above) that could artificially create the experience of religious feelings - the helmet-wearer reported being in the presence of a spirit or having a profound *"feeling of cosmic bliss."*

About eight in every ten volunteers reported quasi-religious feelings when wearing the helmet. However, when an evolutionist and renowned atheist, wore it during the making of a BBC documentary, he famously failed to find God, saying the helmet only affected his breathing and his limbs. Researchers injected radioactive isotope into Buddhists at the point at which they achieved meditative nirvana. Using a special camera, they captured the distribution of the tracer in the brain, which led the researchers to identify the parietal lobes as playing a key role during this transcendental state.

They were more interested in how people coped with everyday moral and religious questions. They said that the latest study suggested the brain was inherently sensitive to believing in almost anything if there were grounds for doing so, but when there was a mystery about something, the same neural machinery was co-opted in the formulation of religious belief.

It seems when we have incomplete knowledge of the world around us, it offers us the opportunities to believe in God. When we don't have a scientific explanation for something, we tend to rely on supernatural explanations. Maybe obeying supernatural forces that we had no knowledge of made it easier for religious forms of belief to emerge.

Are humans hard-wired for faith?

"*I just know God is with me. I can feel Him always,*" a young Haitian woman once said. "*I've meditated and gone to another place I can't describe. Hours felt like mere minutes. It was an indescribable feeling of peace.*"

"*I've spoken in languages I've never learned. It was God speaking through me,*" confided a relative.

The accounts of intense religious and spiritual experiences are topics of fascination for people around the world. It's a mere glimpse into someone's faith and belief system. It's a hint at a person's intense connection with

God, an omniscient being or higher plane. Most people would agree the experience of faith is immeasurable.

Researchers want to change all that. After spending their early medical careers studying how the brain works in neurological and psychiatric conditions such as Alzheimer's and Parkinson's disease, depression and anxiety, they took that brain-scanning technology and turned it toward the spiritual: Franciscan nuns, Tibetan Buddhists, and Pentecostal Christians speaking in tongues. Research team members were surprised by what they found.

"When we think of religious and spiritual beliefs and practices, we see a tremendous similarity across practices and across traditions." Said one researcher, the frontal lobe, the area right behind our foreheads, helps us focus our attention in prayer and meditation.

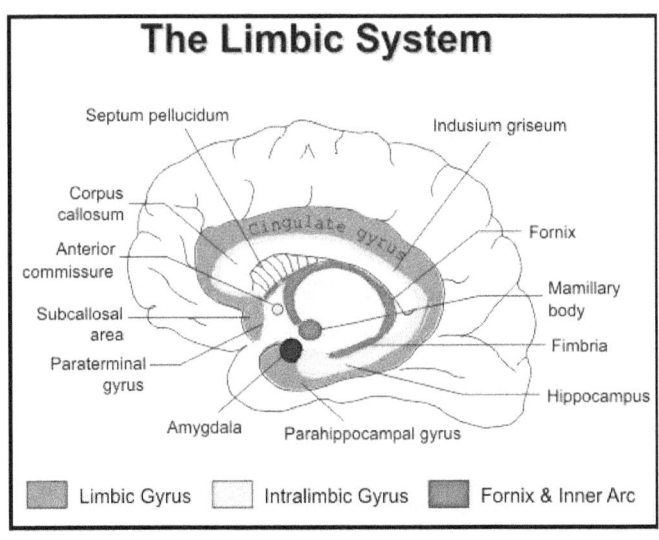

The parietal lobe, located near the backs of our skulls, is the seat of our sensory information

Researchers call the religion the great equalizer and points out those similar areas of the brain are affected during prayer and meditation. They suggest that these brain scans may provide proof that our brains are built to believe in God. There may be universal features of the human mind that actually make it easier for us to believe in a higher power. Interestingly enough, devout believers and atheists alike point to the brain scans as proof of their own ideas.

Some nuns and other believers champion the brain scans as proof of an innate, physical conduit between human beings and God. According to them, it would only make sense that God would give humans a way to communicate with the Almighty through their brain functions.

Some atheists saw these brain scans as proof that the emotions attached to religion and God are nothing more than manifestations of brain circuitry.

Instead of viewing religion and spirituality as an innate quality hardwired by God in the human brain, many of our researchers see religion as a mere byproduct of evolution and Darwinian adaptation. Just like we're not hardwired for boats, but humans in all cultures make boats in pretty much the same way. Now, that's a result both of the way the brain works and of the needs of the world, and of trying to traverse a liquid medium and so we are thinking religion is very much like that.

Researchers point to the palms of his hands as another example of evolutionary coincidence. They say the creases formed there are a mere byproduct of human beings working with our hands -- stretching back to the ages of striking the first fires, hunting the first prey to building early shelter.

Although, the patterns in our palms were coincidentally formed by eons of evolution and survival, he points out that cultures around the world try to find meaning in them through different forms of palm reading. Some Anthropologists say that "*Religion is a byproduct of many different evolutionary functions that organized our brains for day-to-day activity.*"

To be sure, religion has the unparalleled power to bring people into groups. Religion has helped humans survive, adapt and evolve in groups over the ages. It's also helped us learn to cope with death, identify danger and finding mating partners.

Today, scientific images can track our thoughts on God, but it would take a long leap of faith to identify why we think of God in the first place.

A chemical called oxytocin

Researcher's findings found that in terms of numbers roughly six percent of Americans report that they are atheists or agnostics according to a 2012 Pew Research Center poll, leading one to conclude that it must mean over 90 percent believe in a God. Pew also reports that 80 percent profess a religious affiliation and half of those with a religious affiliation regularly attend church. "So *what motivates 120 million Americans to attend a church, synagogue or temple?*"

Researchers began running experiments searching for a biochemical basis for moral behaviors in 2001 and found in a decade's worth of research that the molecule oxytocin (ox - ee – toe – sin) motivated people to return kindness when they were shown kindness during experiments. Researchers only asked college participants the most cursory questions about their religious beliefs. The results were that few college students are religious at all – and they found no difference in oxytocin functioning or pro social behaviors between believers and nonbelievers.

Studies at a Christian church involved taking blood from volunteers before and after a Sunday service and folk dance because many ancient religious rituals - and some current ones - involve coordinated movements. Additionally, samples were taken in rainforest of Papua New Guinea where blood samples from indigenous peoples, during a tribal dance ritual.

In all these tests of blood samples taken during or soon after rituals, a majority of participants released *oxytocin*.

Those who released *oxytocin* also reported that they felt closer to their communities and said they would volunteer to help other community members. They found that the change in oxytocin levels was associated with an increased sense of connection to God or some *"ultimate reality."* Additionally, there was no discernible difference between nonreligious and religious rituals at stimulating oxytocin release. A rugby team that they studied produced as much oxytocin during their warm-up as did evangelical Christians who worshipped and sang in the lab.

Rituals apparently build community by releasing oxytocin. An older woman in their folk-dance sample told them she attended because dancing reduced her chronic pain. I asked her if this was due to the exercise or the people. After a pause, she said "*both.*" Our research data supports her intuition.

Research from other scholars has shown that regular churchgoers are generally happier and healthier. Why? Oxytocin reduces stress responses and having more relationships make us happy.

Churches also give attendants a chance to help others, another activity that increases one's satisfaction with life.

Sure, one researcher says, you don't need God, or church, to be happy or healthy. But the desire to park bottoms in pews has the effects of making people healthy and happy leads churches in the U.S. to be effective community builders. Humans need community and even reminders to be a bit less selfish and help others by studying ancient texts. Perhaps the saying "*God is love- has it backwards*". When we love others, it seems we embody Godliness."

The proportion of atheists and agnostics may keep on growing, but as long as churches use rituals to bring people together, God is unlikely to go away.

In social behavior and wound healing oxytocin is also thought to modulate inflammation by decreasing certain cytokines. Thus, the increased release in oxytocin following positive social interactions has the potential to improve wound healing. One study used heterosexual couples to address this possibility. They found increases in plasma oxytocin following a social interaction were correlated with faster wound healing. They hypothesized this was due to oxytocin reducing inflammation, thus allowing the wound to heal faster. This study provides preliminary evidence that positive social interactions may directly impact aspects of health and this was due to oxytocin reducing inflammation, thus allowing the wound to heal faster. This study provides preliminary evidence that positive social interactions may directly impact aspects of health.

Faith-generated Healing Forces

When doctors try a new drug they generally divide the trial group into two sections. About half the subjects get the new drug and half get sugar pills (made to look like the hopefully-effective product). These inactive tablets are called "placebos".

They do this because patients often experience relief when a doctor or institution in whom they have confidence prescribes anything new and different. In one study of a medication being tried for arthritis, for instance, 40% of the patients who were taking totally ineffective placebo pills experienced substantial relief. Many of them had objective improvement, increased - range of motion, decreased swelling etc. accompanied by diminished discomfort and pain.

Scientists studying what has actually happened in these cases explain that the best explanation seems to be that the patients have developed faith in the "*medication*" (because it was prescribed by someone in whom they had faith) and mobilized confidence-generated healing forces. Many experts are still working understand exactly what those forces are. Scientists just aren't positive whether improvement from placebos has the same basis of facts as those generated by religion-based faith healers, medicine men in primitive tribes, and perhaps some physicians.

Some similar forces also seem able to keep you from getting sick in the first place. A few years ago a team of researchers in Michigan developed a promising cold vaccine. They arranged a substantial trial, dividing their subjects into three groups. One group got the vaccine, one got shots of distilled water (which they were told was vaccine), the third got nothing. Compared with the untreated group those receiving the vaccine got only one-third as many colds in the next season. But the researchers also found that the people who got the distilled water did just as well.

Faith seems even to add to the effect of potent medicines or treatments. Many drug companies today hire people to seek out treatments used by shamans and witch doctors in primitive areas. Often, they find that the herbs, barks and other remedies discovered by trial and error over many generations have chemical can be isolated and manufactured as a drug. In many instances, the effective drug almost never proves quite as effective when administered without ritual as it had been when accompanied by the dances, lying on of hands, or other healer routines.

Religion's Place in Faith Healing

From ancient times, long before medical science achieved substantial cures, most healing efforts were associated with religion. Primitive cultures, for example, generally relied on medicine men, or witch doctors, to pray to the spirits for healing. The ancient Greeks and Romans also prayed to various gods to cure their illnesses and repair the body. For example, infertile couples who wanted to conceive a child would often pray for intervention at the temple of Imhotep in 2700 BC.

Healing by priests or shamans persists in many of the world's current cultures. From **Navajo dances** to **Tibetan prayer-wheels**, the variety of healing efforts based on various non-Christian religions is a staggering amount in the thousands.

One of the largest institutions to practice religion-based faith healing is the **Christian Science organization**. Founded by Mary Baker Eddy, this organization is headed by a charismatic leader who is said to have had experienced an unusual health phenomenon herself:

According to Mary Baker Eddy; *"Believing herself to be pregnant, she had all the usual bodily signs thereof and she ceased to have menstrual periods. She suffered morning sickness for the first two or three months. Her abdomen swelled substantially. But no fetus was present, and after nine months her body resumed its normal non-pregnant state."*

Evidently the extent to which belief could alter body function had something to do with her later career as she became convinced that prayer and bible reading could resolve illnesses. Subsequently she founded an organization which still has thousands of branches throughout the world. Religion based healing is a big business

Although scientists have made phenomenal advances during the twentieth century in understanding and curing sickness and disease, faith healing associated with religion continues to flourish in modern society.

Other examples are the evangelists who claim to heal through prayer and touch. While many believe that these evangelists often stage their miraculous cures, the debate continues to rage over the effectiveness of faith healing. However, the line of demarcation between believers and non-believers is not clear cut.

A 1997 survey of physicians at a meeting of the Academy of Family Physicians found that 99% of them believed that religious faith plays a role in patients' recoveries. Not all of these physicians actually believed in divine intervention but rather that belief can reduce stress and have other psychological effects that help to improve patients' immune system and the ability to fight disease, a belief widespread among scientists and physicians today.

The need to refine and condition the raw natural instincts and desires of individuals in a benign manner is indeed a basic need. Harmonious personal growth is conducive to wholesome interaction with fellowmen, which in turn leads to a salutary impact on the humanity at large.

Many experiments have focused on the role of faith and religion in health. One of the most noted studies took place at the **San Francisco General Medical Center** in 1982 and 1983. The researchers found that the 192 heart patients who were involved with healing prayers of some kind were five times less likely to develop further complications than the 201 who were not. In this study and others, patients were picked at random and not according to their religious beliefs, and they did not know whether or not others were praying them for them.

In 1998, researchers at Duke University also studied 4,000 people over the age of 65. They found that those who participated in religious activities were 40% less likely to have high blood pressure and showed faster recoveries from physical illnesses and depression. Scientific explanations for statistically significant better health and faster recoveries in religious people include healthier lifestyles and a stronger social support that bolsters mental wellbeing.

"One of the saddest lessons of history is this: If we've been bamboozled long enough, we tend to reject any evidence of the bamboozle. We're no longer interested in finding out the truth. The bamboozle has captured us. It's simply too painful to acknowledge, even to ourselves, that we've been taken. **Once you give a charlatan power over you, you almost never get it back.**"

Carl Sagan

We Are All In This Together

Chapter Eleven

The Birth of the Universe

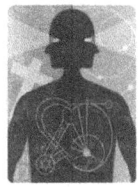

faith engaging
SCIENCE | **RELIGION**

The more we delved into researching both the Biblical and scientific information we realize that the two subjects seldom coincided peaceably. In fact the level of hostility between the two groups of scholars is surprising.

Baruch Spinoza   Stephen Hawking   Albert Einstein

One example of those hostilities was in the 1600's when the Jewish church put a curse on the famous scholar and philosopher **Baruch Spinoza** because of his questionable attitude towards scriptures in the Old Testament. He had developed highly controversial ideas regarding the authenticity of the Hebrew Bible and the nature of the Divine. Questioned by two members of the synagogue, Spinoza at that time apparently responded that

"God has a body and nothing in scripture says otherwise." The Jewish religious authorities issued a cherem (Hebrew: חרם, a kind of ban, shunning, ostracism, expulsion, or excommunication) against him, effectively excluding him from Jewish society at age 23. His books were also later put on the Catholic Church's Index of Forbidden Books.

Another example was the highly publicized conflict between legendary scientist **Albert Einstein** and the church over the years. Albert Einstein's religious views have been studied extensively. He said he believed in the god of Baruch Spinoza, but not in a personal god, a belief he criticized. He also called himself an agnostic, while disassociating himself from the label atheist, preferring, he said an *"attitude of humility corresponding to the weakness of our intellectual understanding of nature and of our own being."*

A modern-day example is the dark reviews by the church received by renowned physicist **Stephen Hawking** when he compared certain religious beliefs in God as tantamount to *"superstitious people who were afraid of the dark."* Hawking argues that invoking God is not necessary to explain the origins of the universe, and that the Big Bang is a consequence of the laws of physics alone. In response to criticism, Hawking has said; *"One can't prove that God doesn't exist, but science makes God unnecessary."*

It all started with a big bang!

Most scientists say that about 13.8 Billion years ago the Universe *"began"* with every speck of its energy jammed into a very tiny point. This extremely dense point exploded with unimaginable force, creating matter and propelling it outward to make the billions of galaxies of our vast universe. Astrophysicists dubbed this gigantic explosion the **Big Bang**.

This extremely volatile tiny speck in the Universe alone didn't seem like such a big deal until something much bigger happened. That something was an even tinier spark that was completely invisible to the naked eye or even the world's most powerful telescopes. The newly discovered / confirmed Higgs Boson aka the *"God particle."*

In May of 2013 at particle accelerators around the world physicists celebrated one of the great moments in scientific history: the discovery of the elusive Higgs boson. Hundreds of physicists and engineers were ecstatic, having devoted almost 30 years of their lives—and $10 billion— trying to track down this almost mythical subatomic particle. In their press release, the scientists at the European Organization for Nuclear Research, better known as CERN, were careful to say they've only found evidence of a *"Higgs-like"* particle. But this is too modest. With 99.9999% confidence, they can claim to have found the Higgs boson itself.

In October of 2013, the discoverers of Higgs boson, a.k.a. "the God particle," were awarded the Nobel Prize in physics. In Stockholm Sweden, **Francois Englert** (left) of Belgium and **Peter Higgs** of Britain won the 2013 Nobel Prize in physics for their theory on how the most basic building blocks of the universe acquire mass, eventually forming the world we know today. The headlines worldwide read *"God particle: Scientists claim building block of universe found - CERN scientists announce observation of Higgs boson particle."*

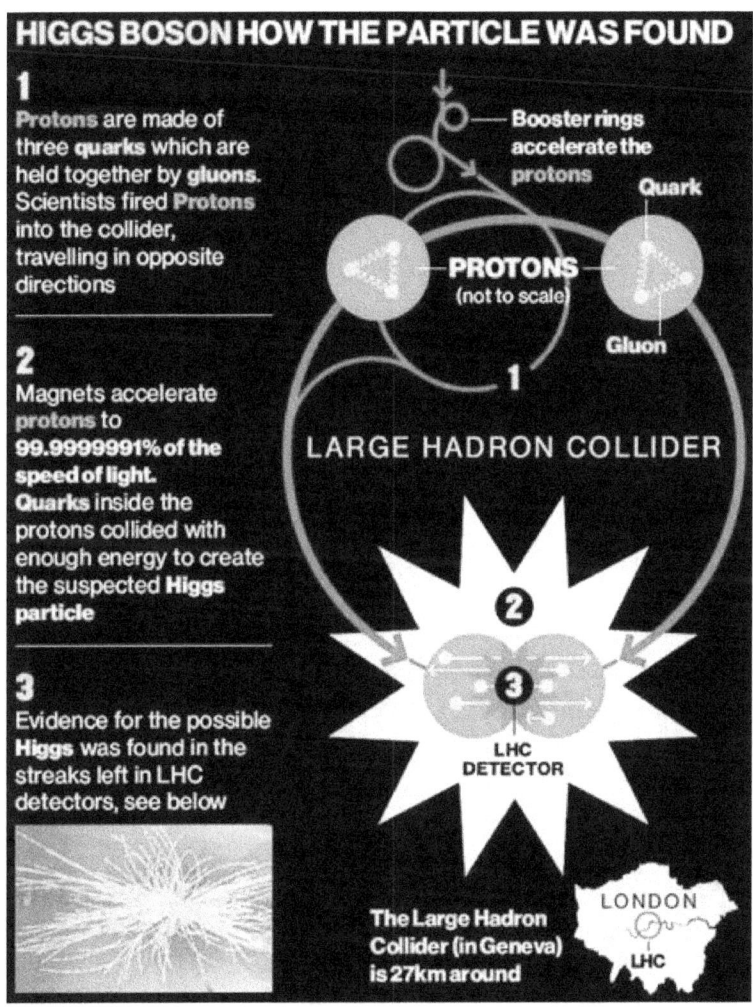

HIGGS BOSON HOW THE PARTICLE WAS FOUND

1

Protons are made of three quarks which are held together by gluons. Scientists fired Protons into the collider, travelling in opposite directions

2

Magnets accelerate protons to 99.9999991% of the speed of light. Quarks inside the protons collided with enough energy to create the suspected Higgs particle

3

Evidence for the possible Higgs was found in the streaks left in LHC detectors, see below

Booster rings accelerate the protons

Quark

PROTONS (not to scale)

Gluon

LARGE HADRON COLLIDER

LHC DETECTOR

The Large Hadron Collider (in Geneva) is 27km around

LONDON

LHC

Their concept was confirmed last year by the discovery of the so-called Higgs particle, also known as the Higgs boson, at CERN, the European Organization for Nuclear Research in Geneva, the Royal Swedish Academy of Sciences said.

"*I am overwhelmed to receive this award and thank the Royal Swedish Academy,*" the 84-year-old Higgs said in a statement released by the University of Edinburgh, where he is a professor emeritus.

"*I hope this recognition of fundamental science will help raise awareness of the value of blue-sky research, of course I'm happy,*" the 80-year-old Englert told reporters, thanking all those who helped him in his research. Asked whether he could have imagined getting a Nobel Prize when he started the research 50 years ago, he said "no *you don't work thinking to get the Nobel Prize, that's not how you work,*" Englert said. "*(Still) we had the impression that we were doing something that was important, that would later on be used by other researchers.*"

Academy member Ulf Danielsson noted that the prize citation also honored the work done at CERN, even though it didn't single out any of its scientists. "*This is a giant discovery, it means the final building block in the so-called standard model for particle physics has been put in place, so it marks a milestone in the history of physics,*" Danielsson said.

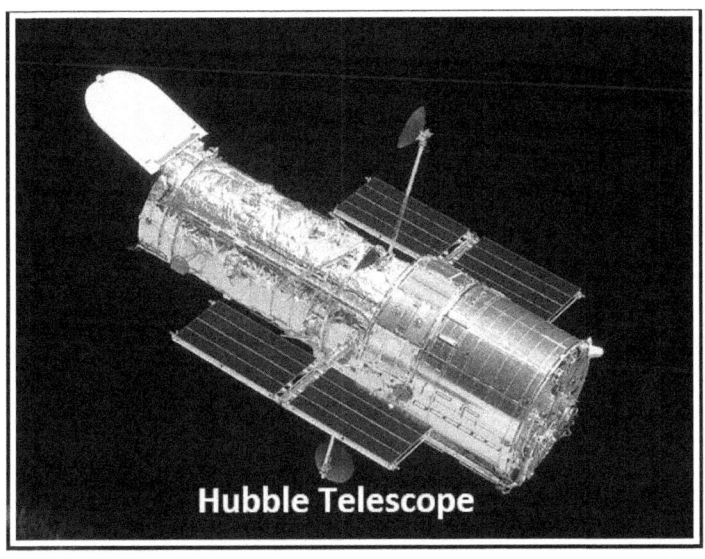

Hubble Telescope

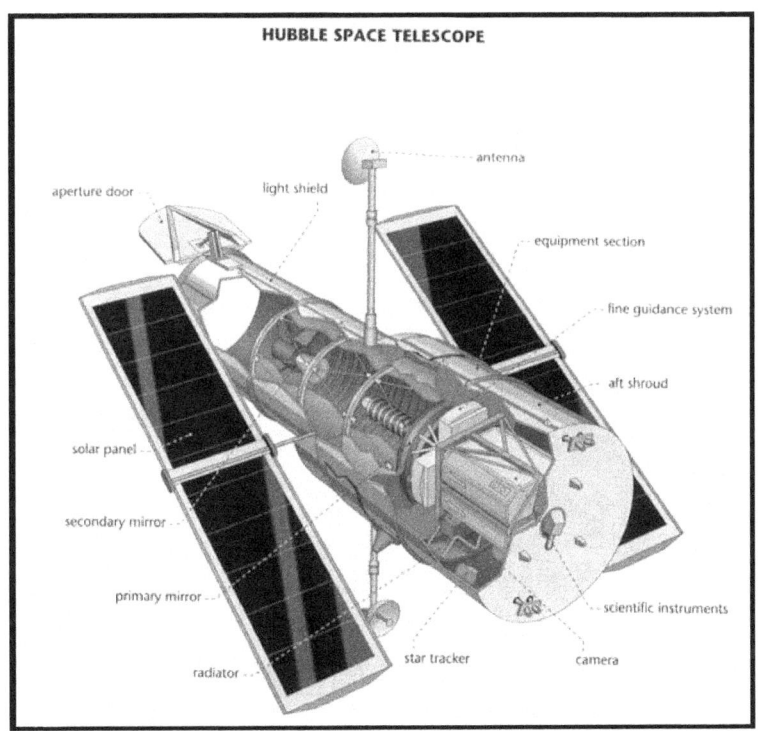

HUBBLE SPACE TELESCOPE

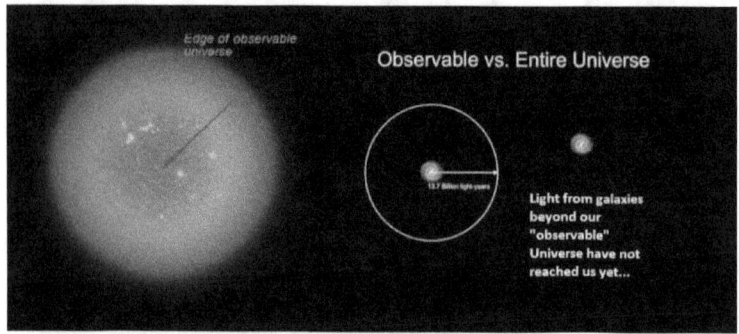

The *"observable"* part of the Universe which is about 93 billion light years in diameter, has led to inferences of its earlier stages. These observations *"suggest"* that the Universe has been governed by the same physical laws and constants throughout most of its extent and history.

The Big Bang theory is the prevailing cosmological model that describes the early development of the Universe, which in physical cosmology is calculated to have occurred approximately 13.798 ± 0.037 billion years ago.

The scientific evidence used to determine the approximate size and age of the universe is based mostly on the following calculations. The speed of light is approximately 186,282 miles per second, or about 5.9 trillion miles per year.

The approximate time that has elapsed since the Big Bang is 13.75 billion years. Multiple the two figures and we find that over the entire history of the universe, light could have travelled approximately 13.75 billion light-years, or 81 billion trillion miles.

However; as stated earlier these calculations have been made based on those parts of the universe that have been "*observable.*" The farthest distance that it is theoretically possible for humans to see is described as the observable Universe.

Observations have shown that the Universe appears to be expanding at an accelerating rate, and a number of models have arisen to predict its size, for example, when the light was first produced.

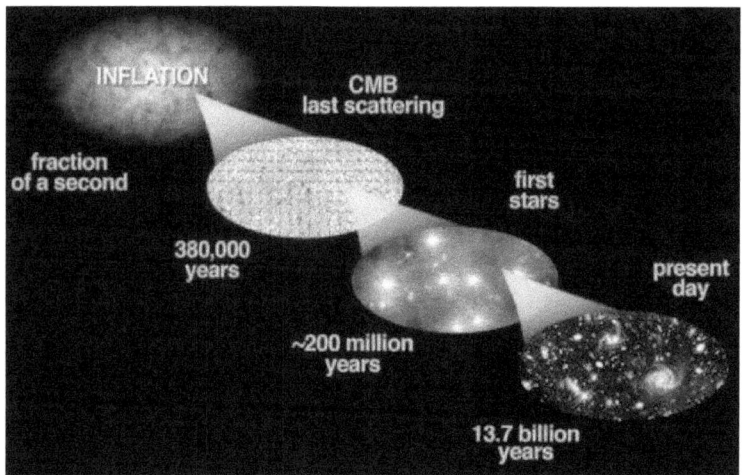

From the time of the Big Bang to the era of "*recombination*" - when neutral hydrogen atoms formed - some 380,000 years later, the universe was "*opaque*" to light. In this case meaning, "*Not able to be seen through; not transparent.*"

Photons bounced between charged particles and didn't travel very far. The reason is that charged particles interact with photons by either absorbing or emitting them. Only *"after"* the era of *"recombination"* could observable light journey through space. That is because photons can pass through neutral hydrogen gas without being diverted. Therefore, any estimate of the size of the observable universe must assume that the farthest light we see was released after that pivotal era when space became transparent.

When the universe was younger than about 380,000 years, the temperature was high enough that all of the hydrogen was ionized, that is the electrons were free and separate from the protons. Because of the presence of the free electrons, photons were scattered around in all directions and could not travel far before changing their direction.

Therefore, the universe was *"opaque."* When the universe cooled to about 3000 K, the electrons and protons combined to form hydrogen atoms. After recombination, photons were able to travel through the universe relatively unimpeded, and the universe became *"transparent."*

The difference between the two times doesn't change the calculation much, but is important to note because as we study *"cause and effects"* regarding the actual formation of the Earth as a habitat for all humanity – being off 380,000 years could become a really big deal.

Another adjustment is far more important. Since the primordial burst of creation, (Big Bang) space has been stretching as the universe expands. A galaxy's distance from us today is far greater than it was when it released the light. The distances traveled by the photons hurled by light sources do not reflect the much greater extent of the sources' current positions. Thus, we could potentially observe light sources that are much farther out than 13.75 billion light-years, if their light was released when they were close enough to Earth.

WMAP aka Wilkinson Microwave Anisotropy Probe

Yet another factor that expands the limit of the observable universe is its acceleration. Not only is the universe expanding; its growth has been speeding up. Data from the Hubble Space Telescope, the WMAP (Wilkinson Microwave Anisotropy Probe) satellite and other instruments have been used to pin down the rate of acceleration, along with the current expansion rate, the age of the universe, and other important cosmological parameters. Researchers, taking advantage of this wealth of information, along with a team of astrophysicists performed a detailed calculation of the radius of the observable universe.

Their answer was 45.7 billion light-years—more than three times bigger than our first, estimate!

Within this sphere lie hundreds of billions of galaxies, each with hundreds of billions of stars. The team calculated this radius by figuring out how far away from us a source would be today if the light we now observe from it was emitted during the recombination era.

Interestingly, as the universe expands, the size of the observable portion will grow—but only up to a point. Researchers showed that eventually there will be a limit to the observable universe's radius: 62 billion light-years.

Because of the accelerating expansion of the universe, galaxies are fleeing from us (and each other) at an ever-hastening pace. Consequently, over time, more and more galaxies will move beyond the observable horizon. Naturally not everything within the observable universe has been identified. It represents the spherical realm that contains all things that could potentially be known through their light signals.

Beyond the observable universe lie *"unknown unknowns"*: the subject of **speculation rather than direct observation**. The 45.7 billion light-year radius includes only light sources. If neutrinos and other particles that could penetrate the opaque conditions of the early universe are included the value becomes **46.6 billion** light-years.

According to most astrophysicists, all the matter found in the universe today -- including the matter in people, plants, animals, the earth, stars, and galaxies -- was created at the very first moment of time. For a brief moment after the Big Bang, the immense heat created conditions unlike any conditions astrophysicists see in the universe today.

While planets and stars today are composed of atoms of elements like hydrogen and silicon, scientists believe the universe back then was too hot for anything other than the most fundamental particles -- such as quarks and photons.

But as the universe quickly expanded, the energy of the Big Bang became more and more diluted in space, causing the universe to cool. Rapid cooling allowed for matter as we know it to form in the universe, although physicists are still trying to figure out exactly how this happened. About one ten-thousandth of a second after the Big Bang, protons and neutrons formed, and within a few minutes these particles stuck together to form atomic nuclei, mostly hydrogen and helium. Hundreds of thousands of years later, electrons stuck to the nuclei to make complete atoms.

COSMOLOGY

Cosmology is the study of the cosmos in several of the above meanings, depending on context. All cosmologies have in common an attempt to understand the implicit order within the whole of being.

Space is a fabric that can bend or twist and is the fabric of the cosmos. With time, it forms a four-dimensional fabric. The fact that the speed of light—the space traveled by light per second—is the same for everyone whether moving or not shows that space-time adjusts itself in a way that light seems to travel at the same speed no matter what.

Space is a flexible fabric that contains the galaxies made of stars and planets as heavy objects. Their weight bends the fabric of space-time which creates a curvature that makes it possible for lighter objects to go around heavier objects, as is the case for **planet Earth** and the lighter weight **moon**, and also for the planets orbiting the sun in our solar system. This phenomenon is known as **gravity** which is one of the four forces of the cosmos.

In simple terms our "*time*" on Earth is based on the twenty-four-hour rotation of the Earth divided into hours, minutes, seconds and microseconds etc. If you want to know what "*time*" it is on Earth you look at your watch or a clock. If you are aboard a space station orbiting Earth and you look at your watch the time will be whatever time it is back on Earth, in any of the 24 time zones, same way with clocks. That's the easy part. The next part is a bit more complicated.

Since space and time are **unified**, the motion through space impacts time: time slows down for the person who is moving but goes faster for the one who stands still; which implies that the passage of time as we experience it may be just an illusion. In that case, every moment in time from the beginning till far in the future coexists together; but in a **different region of the cosmos**. That leads to the concept of time travel: due to the fact that space and time are a unified physical entity it is possible that there are some shortcuts in the fabric of space time that can lead us to another period of time different from our present time.

Despite the possibility of time travel; there is no proof that we can change the past or even the future. The reason: the different periods of time coexist and have a fixed state. Still, the exact nature of time is not fully understood.

Forces of the Cosmos

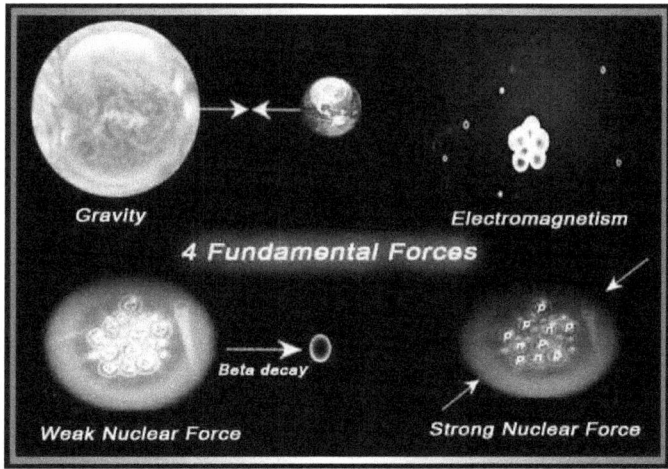

Four Forces of the Cosmos:

- **General relativity:** Einstein's theory of gravity; invoke curvature of space and time.
- **Electromagnetic force:** One of nature's forces; acts on particles that have electric charge.
- **Strong nuclear force:** Force of nature that influences quarks; holds quarks together inside protons and neutrons.
- **Weak nuclear force:** Force of Nature, acting on subatomic scales, and responsible for phenomena such as radioactive decay.

The last 3 forces exist at quantum level which is the atomic scale. The problem is that gravity is the force that happens at the astronomic level and atoms at the quantum level behave differently from the general relativity. With *"gravity"*, the nature of the cosmos is very predictable and ordered, however at the *"quantum level"*, the particle energy and position are unpredictable.

For physicists, it is hard to fit order into the chaos that is happening at atomic level. At the quantum level, gravity force can be explained by the presence of a particle named *"graviton."* This is a particle of energy released by the effect of gravity on the fabric of space-time.

The only issue, no experiment has yet proved the existence of such particle. At the surface space is flexible, ordered but at quantum level it is active, unpredictable, and multidimensional.

The quest for unification

The goal is to combine all the laws in physics into one that can explain the whole cosmos from the outer space down to the quantum scale. Today, there is a popular theory named **String theory** that provides an explanation of the fundamental nature of the Cosmos.

"Superstring Theory" or String theory: Theory in which fundamental ingredients are one-dimensional loops (closed strings) or snippets (open strings), of vibrating energy, which unites general relativity and quantum mechanics.

The fact that the particles behave like waves suggests that they are made of strand shaped like string at the most fundamental level. The shape of the energy strand determines what a proton is and what a neutron is. In other words, different shapes produce different particles. So far, it is just a theory because no scientific experiment *"supports"* or *"disproves"* the string theory.

Most scientists agree with the Big Bang Theory

Science seldom agrees with itself across such a wide spectrum as it does with the Big Bang Theory. There are volumes and volumes of information representing thousands of the world's leading scientists, untold *Billion$* of dollars and vast arrays of scientific equipment including telescopes, satellites, space probes and computers worldwide in agreement.

Particle colliders like the **CERN**, the international space station and many more both classified and unclassified programs from various countries guarantee that science experiments, studies and discoveries have become a daily occurrence in our world today. However; oppositions to the theory that the Earth is a product of the Big Bang – are both few and rare.

How and when the Planet Earth was formed

For as long as humans have occupied the **Earth** – they have been trying to figure out how and when the Earth was made – and by who or what. There have been lots of different ideas about how it happened. Today most scientific research tells us that the Earth was formed with the rest of our solar system about 4.5 billion years ago. Before that the sun, planets and our entire solar system was a lot of space dust. Over millions of years, gravity pulled the dust into a lot of round shapes. One of those was a very early version of our Earth.

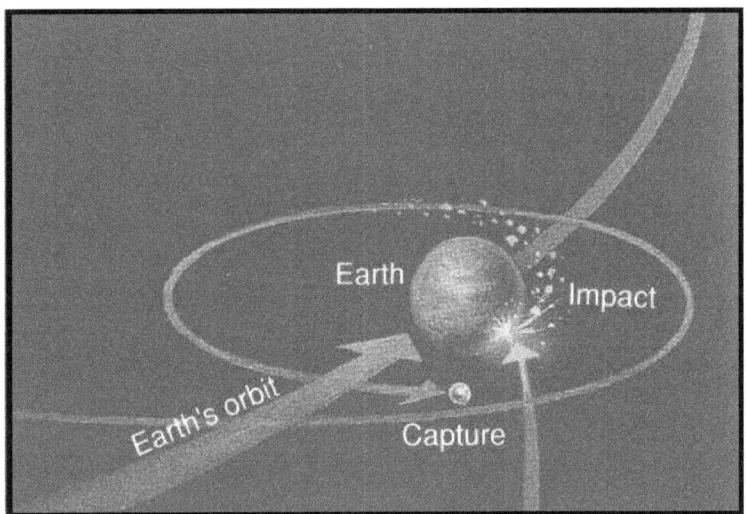

Many scientists think that after becoming round and solid, the **Earth was hit by another object** in our solar system that was about the size of Mars. This object is often called *"Theia"*, and allegedly when it hit the Earth, it knocked out a gigantic chunk of material into space. This material left a huge gap that may have made room for the oceans.

The piece that was knocked into space is most likely now our **moon**.

Moon rocks that were brought back to Earth were date tested by NASA and their age corresponds with that of the Earth.

Orion Nebula

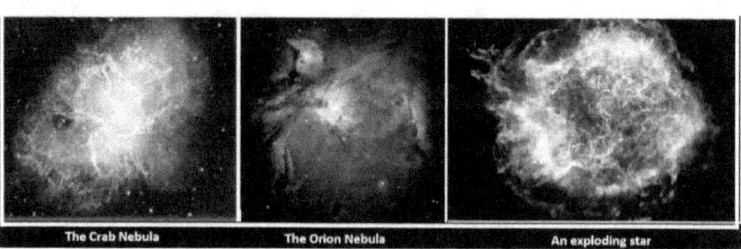

The Crab Nebula   The Orion Nebula   An exploding star

Earth was formed 4.6 billion years ago from a **Nebula** made with dust and gas. All stars and planets are also formed the same way. Earth is still developing the internal core is still in a molten state and causes volcanoes.

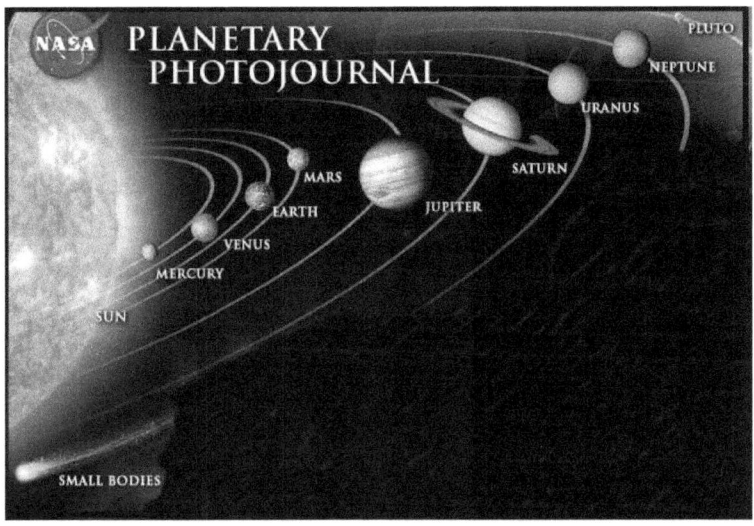

Like Mercury, Venus and the other planets that go around the Sun, the **Earth** was formed from molecules that were floating around space in a nebula cloud of dust that were the leftovers from when the Sun was formed.

At first, the planet forming process began when while flying through space just a few molecules clumped together. Then more and more clumps began crashing into and clinging to each other making even bigger clumps. After a while there were much bigger collisions as pieces of metal and rock – some hundreds of miles across - smashed into and clung to each other. The process continued until millions of years later the mass was finally big enough to become a planet. The Sun's gravity caught and held the newly formed planet Earth (along with other new planets like Mercury and Mars) as they maintain a continued orbit around the Sun.

The Earth's orbiting momentum is always trying to pull it away from the Sun in a straight line while the Sun's gravity pulls it back causing the centrifugal force that keeps the Earth in a continual orbit of about 92 million miles, (150 million kilometers. Nowadays it takes the Earth one year to go all the way around the Sun (moving at about 67,000 miles per hour, or 100,000 kilometers per hour. This annual orbit of the Earth around the Sun is what causes the seasons to change: spring, summer, fall, and winter.

Most theories agree that Earth was formed from cast off material from a star.

The Big Bang theory holds that all matter came from one great explosion of matter and another theory holds that the bang came from the end of the collapse of a black hole when the energy absorbed became greater than the gravity of the hole. Yet another theory contends that the universe is forever expanding then falling in upon itself, repeatedly throwing out stars and planets, then drawing them back in to the center. It's like a never-ending process.

Every culture has its own mythology concerning the origin of the universe and the formation of the Earth. When you read them all, though, they bear striking similarities to each other and to modern scientific theory. *At the moment, science for the most part is agreeing that with the theory that the Higgs Boson aka the "God Particle," was the spark that caused the Big Bang*

A little less than five billion years ago, a star became super nova and exploded, causing a tidal wave of matter. Some of it started whirling around, smashing together to form bodies of burning hydrogen and helium. A large body, our Sun, had enough mass to exert enough gravity to gather gaseous clouds of matter in orbit around it. Matter in this cloud combined to form smaller bodies which, as they orbited the Sun, smashed into each other, forming larger bodies which became planets, asteroids and other planetisimals and cometisimals (Refers to *"Any of numerous small celestial bodies that may have existed at an early stage of the development of the solar system."*)

The third larger body circling the sun, proto earth, was a ball of boiling molten rock. As it cooled, though, the denser metals sank toward the center and the lighter metals surfaced and formed a crust. As the planet cooled, gas escaped from the center, throwing up volcanoes on the surface that belched water vapor, carbon dioxide, sulfur, nitrogen, argon and chlorine up.

A few planetary collisions later, proto earth had enough water vapor in orbit around it to begin forming an atmosphere. If the atmosphere of Venus and Mars are any indication, Earth had the right planetary mass and was far enough away from the sun to disperse most of the carbon dioxide in its atmosphere but hold on to a hospitable atmosphere composed of 75 percent nitrogen and 25 percent oxygen.

More planetary collisions occurred as Earth rolled on, contributing mass, water vapor and a moon to its progress. When enough water vapor accumulated in the atmosphere, it began to rain, further cooling the surface. Running water formed channels and oceans as the crust split and re-formed giant continents of solid crust floating on the shifting mantle over the molten core. All of the elements were present, whether by cosmic accident or some design, for life to begin developing on the new planet we call Earth.

The formation processes

About 4.5 billion years ago - in the beginning - the Earth was very different from what it is now. There was no atmosphere and no water bodies on the surface of the Earth.

It is widely accepted by research scientists that the sun, the planets and their satellites in our solar system were all formed out of solar nebula. (Refers to) *"Solar nebula as a huge mass of spiral cloud made up of dust particles and various types of hot gases."*

The key constituent elements of this solar nebula were hydrogen and helium gases and there were some other heavier elements too. The Solar nebula began its contraction about 4.6 billion years ago probably because of shock wave caused by stellar explosion in the nearby region. During the contraction process, its temperature came down and it began to rotate very fast.

As a result of cooling, shrinking and faster rotation, the outer part of the cloud got detached from the main body. The solar system was formed out of one of these detached rings of the gas cloud. The ring or disk shrank further and the center of the dense cloud heated up to form the sun, while the outer parts of the ring clustered together due to gravity to become proto planets. One such cluster of gases created the planet Earth.

This ball of hot gases was by now, cooled so much that all the gases in it were condensed into liquid or lava. Over a period of time, with further decrease in temperature, the lava solidified and led to the formation of the crust of the Earth. During the process of solidification, the liquid heavy metals like iron, descended to the center due to high density, forming the core of the Earth. The remaining part of the lava formed the Earth's mantle, which lies just below the crust.

Due to the cooling of Earth, large amount of steam escaped from its crust. As a result of the eruptions of volcanoes, substantial amounts of steam and various types of gases were also released. Geologists believe that a good amount of water was imported to the Earth during its collisions with several comets that contained ice. When the temperature on the Earth was very high, all the water remained in vaporous state and surrounded the surface of the Earth. Then, as the temperature on Earth decreased, all the steam condensed to form clouds. The resultant rainwater got accumulated in the craters formed on the

Earth's surface by the impact of smashing of comets. This led to the formation of the oceans.

In the early stage, the Earth's atmosphere composed of hydrogen and helium gases. However, as the Earth's gravity was not that strong then, it could not hold on to these light gases and they were lost in the space. Later on, the impact of collisions with comets enriched the Earth with water as well as many other essential gases like carbon dioxide, methane, nitrogen, ammonia, etc. The gases discharged from volcanic eruptions and collisions with comets also contributed to the formation of the new atmosphere over the Earth. Free oxygen was not available in the new atmosphere; it was either bound by hydrogen or some other elements. The ozone layer was absent in the newly formed atmosphere. As a result, the Earth's surface was exposed to ultraviolet radiation. Significant amount of free oxygen was found in the atmosphere only after life forms came to the Earth when the photosynthesis process supplied oxygen to the atmosphere.

According to theories of geology, the present-day continents were formed due to fragmentation of a huge, single mass of land. The Earth's crust comprises a number of large plates of solid rock floating on the liquid mantle. The molten rock of the Earth's mantle is in constant motion due to convection of heat that occurs deep inside it. Due to this internal motion, some of the plates are constantly sliding at the edges (more commonly) in relation to others. This kind of continuous movement detached one plate from other. The broken pieces of land masses then drifted away from each other and caused the division of the continents. It is believed that the mountains were formed when one plate of the crust pushed itself against another and the resultant pressure thrust a part of the land upwards. All the information about how was the Earth formed provided in this article is based upon various research studies conducted by geologists, cosmologists, etc.

All of these changes took place on the planet gradually over a period of several millions of years. The geological features of the Earth then provided a suitable condition for evolution of life on the planet which appeared within one billion years of its formation. However, the human species came into existence much later.

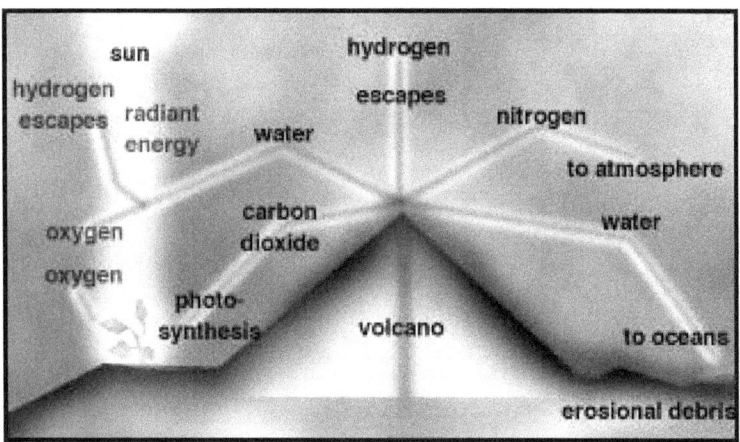

The atmosphere is the skin of air around the earth. It's gone through several different stages since the earth formed.

The quick version is that the first atmosphere came from **volcanoes** and was mostly water and carbon dioxide. As it cooled down it rained and made the oceans and a lot of the carbon dioxide dissolved. Later some types of algae started making oxygen, until eventually the atmosphere was like it is today.

When the earth formed it was so hot the rock was molten, like lava all over. There were lots of chemical reactions that made gases, but they were stripped away by the strong wind that comes out of the sun (solar wind) very quickly. When earth cooled a bit it formed a crust but there were still lots of volcanoes, gases came from them to form the first atmosphere. It was mostly water – the earth was so hot there were no oceans and the water was all steam and clouds.

As **Earth** cooled more there was the **first storm** – all the steam in

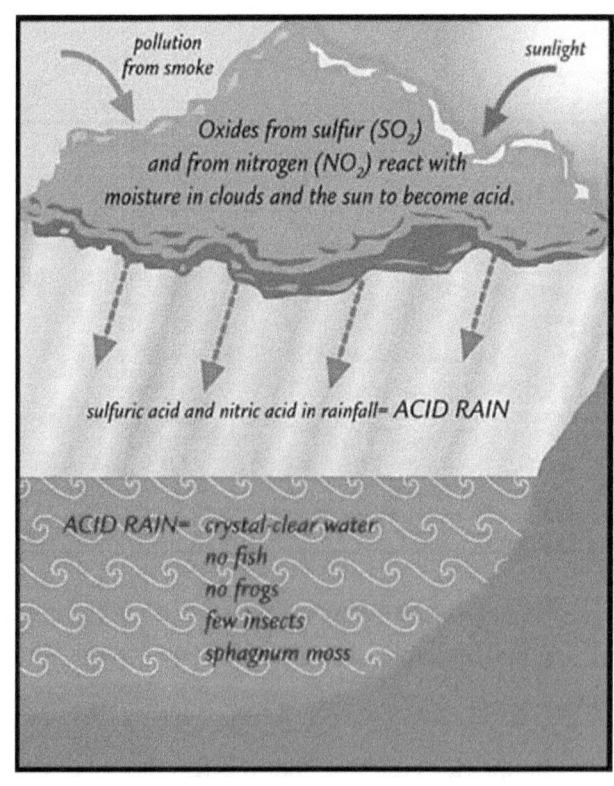

pollution from smoke

sunlight

Oxides from sulfur (SO_2) and from nitrogen (NO_2) react with moisture in clouds and the sun to become acid.

sulfuric acid and nitric acid in rainfall= ACID RAIN

ACID RAIN= crystal-clear water, no fish no frogs few insects sphagnum moss the atmosphere came down in a massive rainfall and formed the oceans, and the gases that were left made the second atmosphere, along with more gases from volcanoes. There was a lot more carbon dioxide than today, but once the oceans formed it started dissolving in them. This all took a very long time – hundreds of thousands of years.

Somewhere in the oceans life began. Early life didn't use oxygen because there wasn't any; it breathed other gases like methane. But some algae made oxygen, the same way plants make it today. At first it only built up very, very slowly, but oxygen was poisonous to most of the things that were alive. When there was enough of it the oxygen started to kill most things, except a few that could tolerate it and even use it. When these were just about the only living things left they spread very quickly and started making even more oxygen.

Once there was oxygen the ozone layer could form, because ozone is a type of oxygen. It is up high above the earth and protects us from ultra-violet (UV) radiation from the sun – the one that burns you. The reason being sunburnt is dangerous is that UV radiation doesn't just burn your skin; it can damage the DNA in your skin cells and turn them into cancers. Until the ozone layer formed nothing could live on land because the UV would kill them. But once it was there life could move out of the oceans.

The atmosphere is constantly changing. Today we are putting artificial chemicals into it as air pollution, many of these have had unexpected effects. We are also increasing the carbon dioxide which is making the earth warm up.

But even without us, volcanoes, fires, the way the continents move, the sun and of course other living things are all changing the atmosphere all the time.

Earth is the only planet we know of that can support life as we know it on Earth. This is an amazing fact, considering that it is made out of the same matter as other planets in our solar system, was formed at the same time and through the same processes as every other planet, and gets its energy from the sun.

To a universal traveler, Earth may seem to be a harmless little planet in the far reaches of one of billions of spiral

galaxies in the universe. It has an average size star of average brightness and is joined by seven other planets — which support no known life forms — in its solar system.

While this may be fitting for a passage from The Hitchhikers Guide to the Galaxy, by Douglas Adams, in the grand scheme of the universe, it would be a fairly accurate description.

However, Earth is a planet teeming with vitality and is home to - billions of plants and animals that share a common evolutionary track. How and why did we get here? What processes had to take place for this to happen? And where do we go from here? The fact is, no one has been able to come close to knowing exactly what led to the origins of life, and we may never know. After 5 billion years of Earth's formation and evolution, the evidence may have been lost. But scientists have made significant progress in understanding what many processes that may have led to the origins of life.

Scientists studying the *"origins of life"* are looking very hard nowadays at certain elements of our Universe that not necessarily *"could"* have supplied the building blocks of life – but *did.* It seems at every turn or up every hill that new information surfaces that researchers are amazed at. Especially when that amazement involves many formulas and laws of physics that deal with astronomically huge numbers; yet apparently against all odds living forces continue to prevail. And this hasn't just happened in a few isolated cases – it has happened millions of times.

Simple as it may sound it's kind of like repeatedly throwing the dice and they keep coming up seven! There is such a thing as a lucky streak – but many scientists are convinced that maybe it's all because of something else, perhaps a *"fine-tuned Universe."*

Chapter Twelve

Is this a fine-tuned Universe?

Is this a fine-tuned Universe?

The fine-tuned Universe theory is the proposition that the conditions that allow life in the Universe can only occur when certain universal fundamental physical constants lie within a very narrow range, that means if any of several fundamental constants were only *slightly different*, the Universe would be unlikely to be conducive to the establishment and development of matter, astronomical structures, elemental diversity, or simply put – **support life** as it is presently understood.

The existence and extent of fine-tuning in the Universe is a *matter of dispute* in the scientific community. The proposition is also discussed among philosophers, theologians, creationists, and intelligent design proponents.

Modern scientists assert that "*There is now broad agreement among physicists and cosmologists that the Universe is in several respects 'fine-tuned' for life*". However, they continue, "*the conclusion is not so much that the Universe is fine-tuned for life; rather it is fine-*

tuned for the building blocks and environments that life requires." They also state that" *'anthropic' reasoning fails to distinguish between minimally biophilic universes, in which life is permitted, but only marginally possible, and optimally biophilic universes, in which life flourishes because biogenesis occurs frequently"*. Among scientists who find the evidence persuasive, a variety of natural explanations have been proposed, such as the anthropic principle along with multiple universes.

In astrophysics and cosmology, the anthropic principle *"is the philosophical consideration that observations of the physical Universe must be compatible with the conscious life that observes it."* And the *biophilia hypothesis* *"suggests that there is an instinctive bond between human beings and other living systems"*. Biophilia is defined as *"the urge to affiliate with other forms of life."*

Early in scientific history one of the first studies to explore concepts of fine tuning in the Universe discusses the importance of water and the environment with respect to living things, pointing out that life depends entirely on the very specific environmental conditions on Earth, especially with regard to the prevalence and properties of water.

Additionally, they claimed that certain forces in physics, such as *gravity* and *electromagnetism*, must be perfectly fine-tuned for life to exist *anywhere* in the Universe. One scientist compared *"the chance of obtaining even a single functioning protein by chance combination of amino acids to a star system full of blind men solving Rubik's Cube simultaneously."*

Today it is widely thought that carbon-based life was not haphazardly arrived at, but the deliberate end of a Universe *"tailor-made for man."*

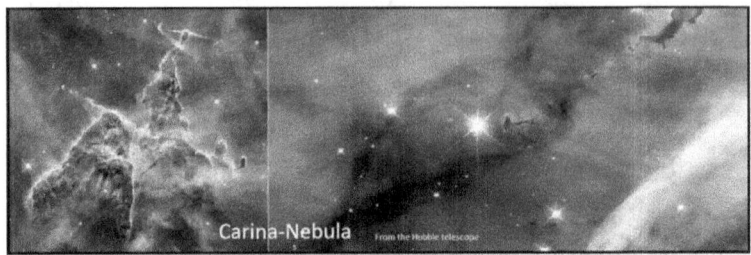

Carina-Nebula From the Hubble telescope

Fine-tuned Universe proponents argue that deep-space structures such as the **Carina Nebula** would not form in a universe with *minimally different* physical constants.

The premise of the fine-tuned Universe assertion is that a small change in several of the dimensionless fundamental physical constants would make the Universe radically different.

If, for example, the strong nuclear force were 2% stronger than it is or if the coupling constant representing its strength were 2% larger, while the other constants were left unchanged, diprotons would be stable and hydrogen would fuse into them instead of deuterium and helium. This would drastically alter the physics of stars, and presumably preclude the existence of life similar to what we observe on Earth.

The existence of the *di-proton* would short-circuit the slow fusion of hydrogen into deuterium. Hydrogen would fuse so easily that it is likely that all of the Universe's hydrogen would be consumed in the first few minutes after the Big Bang. However, some of the fundamental constants describe the properties of the unstable strange, charmed, bottom and top quarks and mu and tau leptons that seem to play little part in the Universe or the structure of matter.

The **precise formulation** of the idea is made difficult by the fact that physicists do not yet know how many independent physical constants there are. The current standard model of particle physics has 25 freely adjustable parameters with an additional parameter, the cosmological constant, for gravitation. However, because the standard model is not mathematically self-consistent under certain conditions (e.g., at very high energies, at which both quantum mechanics and general relativity are relevant), physicists believe that it is under laid by some other theory, such as a grand unified theory, string theory, or loop quantum gravity.

In some candidate theories, the actual number of independent physical constants may be as small as one. For example, the cosmological constant may be a fundamental constant, but attempts have also been made to calculate it from other constants, and according to the author of one such calculation, *"the small value of the cosmological constant is telling us that a remarkably precise and totally unexpected relation exists among all the parameters of the Standard Model of particle physics, the bare cosmological constant and unknown physics."*

Computer simulations suggest that not all of the purportedly *"fine-tuned"* parameters may be as fine-tuned as has been claimed. Researchers have simulated different universes in which four fundamental parameters are varied. He found that long-lived stars could exist over a wide parameter range, and   concluded that "*a wide variation of constants of physics leads to universes that are long-lived enough for life to evolve, although human life need not exist in such universes.*"

A later study investigating the structure of stars in universes with different values of the gravitational constant G, the fine-structure constant α, and a nuclear reaction rate parameter C. study suggests that roughly 25% of this parameter space allows stars to exist.

The validity of fine tuning examples is sometimes questioned on the grounds that such reasoning is subjective "*anthropomorphism*;" (the assignment of human attributes to nonhuman things) applied to natural physical constants. Critics also suggest that the fine-tuned Universe assertion and the anthropic principle are essentially tautologies – the saying of the same thing twice in different words. The fine-tuned Universe argument has also been criticized as an argument by lack of imagination, as it assumes no other forms of life, sometimes referred to as **carbon chauvinism**.

Conceptually, alternative biochemistry or other forms of life are possible. In addition, critics argue that humans are adapted to the Universe through the process of evolution, rather than the Universe being adapted to humans.

Many also see it as an example of the logical flaw of in its assertion that humans are the purpose of the Universe. Simply put *"arrogantly assuming that human beings are the most significant **entity** of the Universe reminds us that it's all a matter of perspective isn't it."*

Possible naturalistic explanations

There are fine tuning arguments that are naturalistic. As modern cosmology developed, various hypotheses have been proposed. One is an oscillatory universe or a multiverse, where fundamental physical constants are postulated to resolve themselves to random values in different iterations of reality. Under this hypothesis, separate parts of reality would have wildly different characteristics.

In such scenarios, the issue of fine-tuning does not arise at all, as only those *"universes"* with constants hospitable to life (such as what we observe) would develop life capable of contemplating the question of the origin of fine-tuning.

Based upon the Anthropic principle some argue that the structure (age, physical constants, etc.) of the Universe as seen by living observers is not random, but is constrained by biological factors that require it to be life friendly.

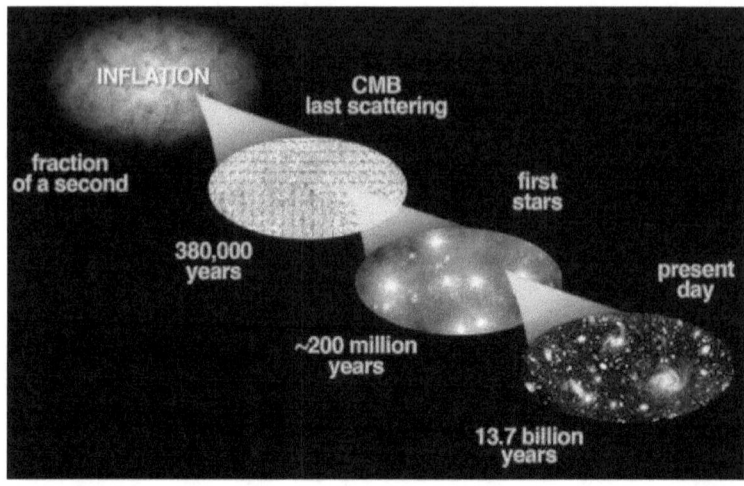

Inflation theory posits that an inflation field in the first 10−30 seconds of the universe produces strong repulsive gravity, and the universe and space-time expand by a factor of 1030. After 10−30seconds, gravity starts to become attractive. In this framework, with such rapid expansion, the overall shape of the universe at 14 billion years is much less sensitive to initial parameters than the standard big bang model, and thus the fine-tuning issue disappears.

The Anthropic Coincidences

In standard inflation, inflationary expansion occurred while the Universe was in a false vacuum state, halting when the Universe decayed to a true vacuum state. The bubble universe model proposes that different parts of this inflationary universe (termed a Multiverse) decayed at different times, with decaying regions corresponding to universes not in causal contact with each other. It further supposes that each bubble universe may have different physical constants.

Scientists of CERN, proposed that the Universe's initial conditions consisted of a superposition of many possible initial conditions, only a small fraction of which contributed to the conditions we see today. According to their theory, it is inevitable that we find our Universe's "*fine-tuned*" physical constants, as the current Universe "*selects*" only those past histories that led to the present conditions. In this way, top-down cosmology provides an anthropic explanation for why we find ourselves in a universe that allows matter and life, without invoking the existence of the Multiverse.

Multiverse

The Multiverse hypothesis assumes the existence of many universes with different physical constants, some of which are hospitable to intelligent life (see multiverse: anthropic principle). Because we are intelligent beings, we are by definition in a hospitable one. Researchers have argued that the anthropic principle and selection effect resolves the entire issue of fine-tuning, while another reaches the opposite conclusion using Bayesian probability.

Anthropic Principle

This approach has led to considerable research into the *anthropic principle* and has been of particular interest to particle physicists, because theories of everything do apparently generate large numbers of universes in which the physical constants vary widely. As yet, there is no evidence for the existence of a multiverse, but some versions of the theory do make predictions that some researchers studying M-theory and gravity leaks hope to see some evidence of soon. Some multiverse theories are no falsifiable, thus scientists may be reluctant to call any multiverse theory "*scientific*".

Variants on this approach include one scientist's notion of cosmological natural selection, the Ekpyrotic universe, and the Bubble universe theory. Critics of the multiverse-related explanations argue that there is no evidence that other universes exist.

Bubble universe theory

The bubble universe model postulates that our Universe is one of many that grew from a multiverse consisting of vacuum that had not yet decayed to its ground state. According to this scenario, by means of a random quantum fluctuation, the Universe "*tunneled*" from pure vacuum ("*nothing*") to what is called a false vacuum, a region of space that contains no matter or radiation, but is not quite "*nothing*." The space inside this bubble of false vacuum was curved, or warped. A small amount of energy was contained in that curvature, somewhat like the energy stored in a strung bow.

This ostensible violation of energy conservation is allowed by the **Heisenberg uncertainty principle** for sufficiently small-time intervals. The bubble then inflated exponentially and the Universe grew by many orders of magnitude in a tiny fraction of a second. As the bubble expanded, its curvature energy was converted into matter and radiation, inflation stopped, and the more linear Big Bang expansion we now experience commenced. The Universe cooled and its structure spontaneously froze out, as formless water vapor freezes into snowflakes whose unique patterns arise from a combination of symmetry and randomness.

One hypothesis is that the Universe may have been designed by ***extra-universal aliens.*** Some believe this would solve the problem of how a designer or design team capable of fine-tuning the Universe could come to exist. Cosmologist Alan Guth believes humans will in time be able to generate new universes. By implication previous intelligent entities may have generated our Universe. This idea leads to the possibility that the extraterrestrial designer/designers are themselves the product of an evolutionary process in their own universe, which must therefore itself be able to sustain life. However, it also raises the question of where this universe came from, leading to an infinite regress.

The Simulation hypothesis promoted by Nick Bostrom and others suggests that our Universe may be a computer simulation by aliens.

The Biocosm hypothesis and the **Meduso-anthropic principle** *both* suggest that natural selection has made the universe biophilic. The Universe enables intelligence because intelligent entities later create new biophilic universes. This is different from the suggestion above that aliens from a universe that is less-finely tuned than ours made our Universe finely tuned. **The Designer Universe theory** suggests that the Universe was designed by members of a technologically advanced civilization in another part of the multiverse, and that this advanced civilization may have been responsible for causing the Big Bang.

Religious arguments

As with theistic evolution, some individual scientists, theologians, and philosophers as well as certain religious groups argue that providence aka (under the guidance or care of God) or creation are responsible for fine-tuning.

One Christian philosopher, argues that

random chance, applied to a single and sole universe, only raises the question as to why this universe could be so "*lucky*" as to have precise conditions that support life at least at some place (the Earth) and time (within millions of years of the present).

One reaction to these apparent enormous coincidences is to see them as substantiating the theistic claim that the Universe has been created by a personal God and as offering the material for a properly restrained theistic argument—hence the fine-tuning argument. It's as if there are a large number of dials that have to be tuned to within extremely narrow limits for life to be possible in our Universe. It is extremely unlikely that this should happen by chance, but much more likely that this should happen, if there is such a person as God.

This fine-tuning of the Universe is cited by yet another theologian and philosopher as an evidence for the existence of God or some form of intelligence capable of manipulating (or designing) the basic physics that governs the Universe. They argue, however, "*that the postulate of a divine Designer does not settle for us the religious question.*"

Intelligent design

Proponents of Intelligent design argue that certain features of the Universe and of living things are *best explained* by an intelligent cause, not an *undirected* process such as *natural selection*. The fine-tuned Universe argument is a central premise or presented as given in many of the published works of prominent intelligent design proponents.

The Counter argument to religious views argues that "*The fine-tuning argument and other recent intelligent design arguments are modern versions of God-of-the-gaps reasoning, where a God is deemed necessary whenever science has not fully explained some phenomenon.*"

The argument from imperfection suggests that if the Universe were designed to be fine-tuned for life, it should be the best one possible and that evidence suggests that it is not. In fact, most of the Universe is highly hostile to life.

"We have no reason to believe that our kind of carbon-based life is all that is possible. Furthermore, modern cosmology indicates that multiple universes may exist with different constants and laws of physics. So, it is not surprising that we live in the one suited for us.

The Universe is not fine-tuned to life; apparently life is fine-tuned to the Universe."

Chapter Thirteen

What really happened to the Dinosaurs?

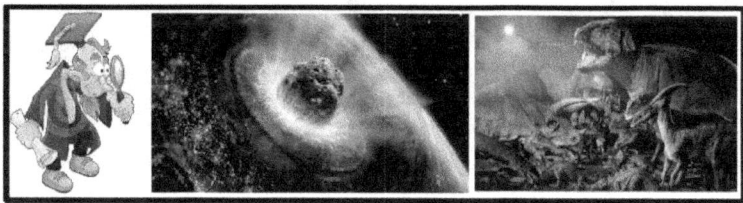

What really killed the Dinosaurs?

Our quest for knowledge has led us on a scientific steeple chase discovery mission like we have never been on before. Just in the last year alone we have learned about the discovery of the Higg's Boson aka the "*God Particle*," and its confirmation that it's "*spark*" caused the Big Bang. We learned of evidence that now says our Universe is probably 4-5 times larger than previously thought. Additionally, science is now becoming able to reverse engineer carbon based life forms and have discovered life elsewhere in the Universe that is nothing like us or anything else on earth.

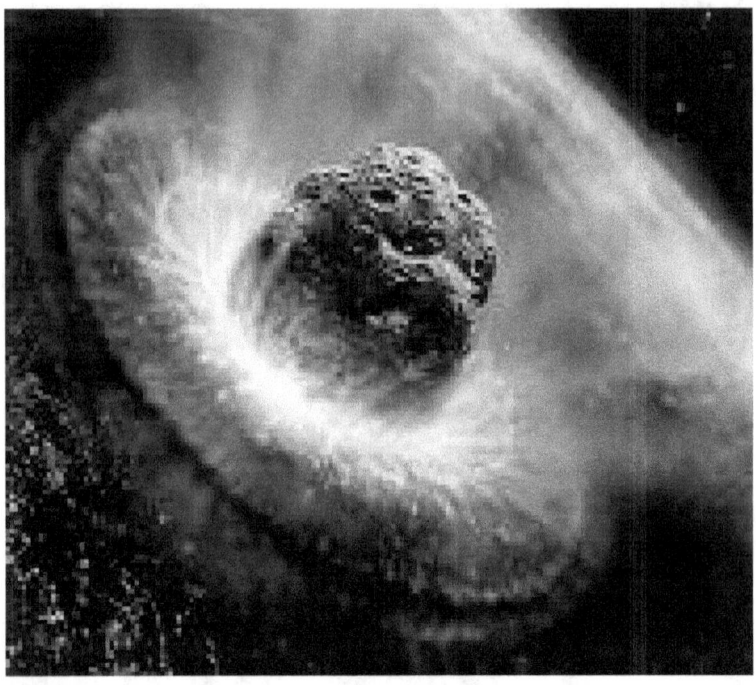

And if that isn't enough to dazzle our star struck minds – now we are learning that *maybe* the "*historic* **asteroid**" that killed off the dinosaurs – wasn't the lone culprit after all! In fact, it appears "***many***" events ranging from global cooling, forest fires, acid rain, smoke filled atmosphere, 24 hour a day darkness and DNA mutation caused by the Sun's ultraviolet rays – all probably contributed to their extinction. But more important than an interesting history lesson, this new information may be the beginning of how to get Earth back on an environmentally correct path to give our future ancestors a chance (like us) to survive.

Yucatan peninsula off the coast of Mexico

Let's examine a few existing theories. For example: Did a collision with a giant asteroid or comet change the shape of life on Earth forever? It is widely agreed that such an -- 10 kilometer across -- struck just off the coast of the Yucatan peninsula, Mexico about 65 million years ago. According to some scientists who maintain that dinosaur extinction came *quickly*, the impact must have spelled the cataclysmic end for the dinosaurs but paved the way for the development of the Human race as we've come to know it.

Some scientists have concluded that dense clouds of dust blocked the sun's rays, darkening and chilling Earth to deadly levels for most plants and, in turn, many animals. Then, when the dust finally settled, leaving greenhouse gases created by the impact near the Cretaceous/Tertiary (KT) boundary, the geological layer began to define the dinosaur extinction.

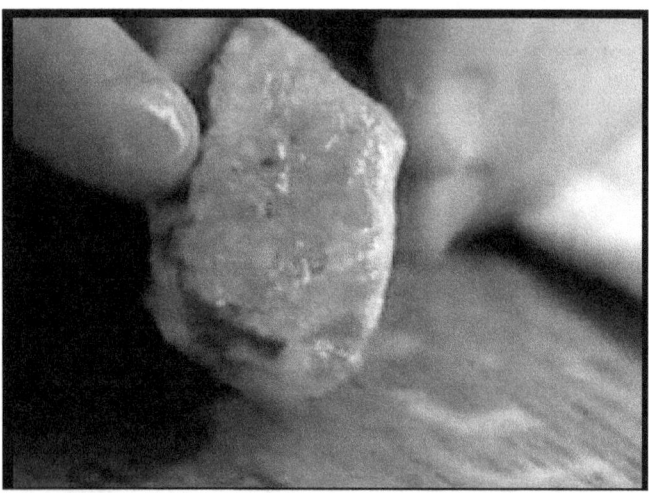

Fractured Crystals (Shocked Quartz) are yet another tell-tale trademark typically found in areas of suspected asteroid collisions with Earth.

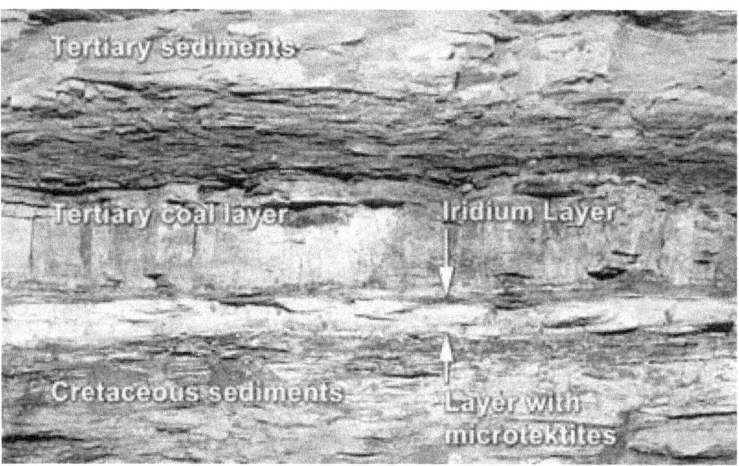

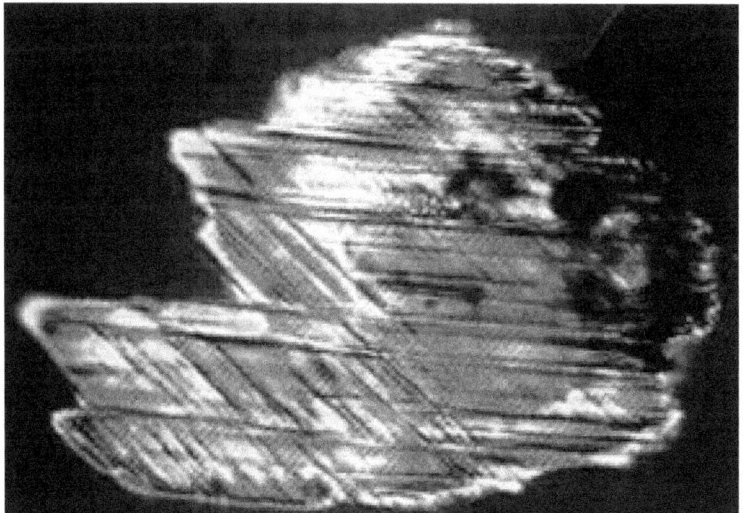

These crystals, often called *"shocked quartz"* show a **distinctive pattern** of fracturing caused by high-energy impacts or explosions. Some scientists maintain that the fracture pattern in these quartz crystals could *only* have been caused by a massive asteroid or comet impact.

The pattern is prevalent in quartz found at or near the Cretaceous/Tertiary (KT) boundary - the geological layer deposited at the time of the extinction.

A gradual decline in the number of **dinosaur species** would likely mirror an equally gradual cause of their ultimate extinction.

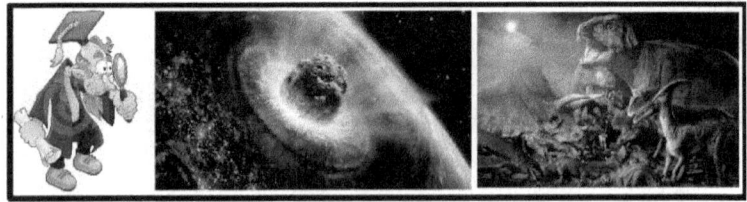

Conversely, a sudden "*here today – gone tomorrow*" type end to the dinosaurs implies what could be a *catastrophic* cause.

Depending on location and interpretation, the fossil record seems to say many different things. While some scientists say that this asteroid's high energy impact caused the extinction of the dinosaurs other scientists and paleontologists see evidence in the fossil record that dinosaurs were doing quite well prior to the end of the Cretaceous period and that they were in no way declining in abundance when the impact occurred.

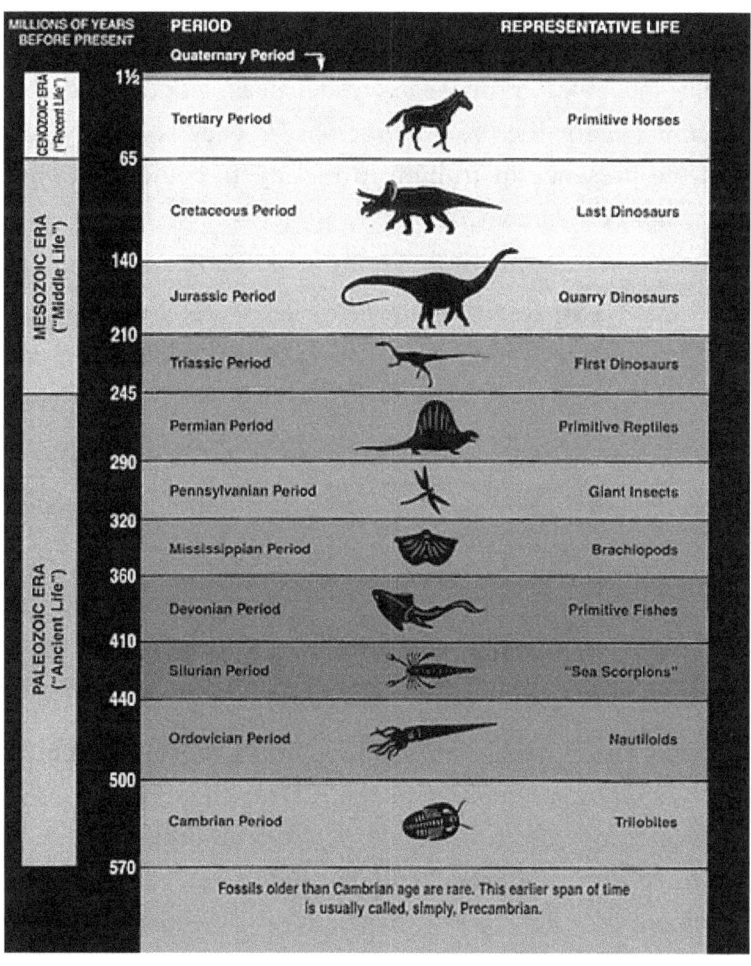

MILLIONS OF YEARS BEFORE PRESENT		PERIOD	REPRESENTATIVE LIFE
		Quaternary Period	
CENOZOIC ERA ("Recent Life")	1½	Tertiary Period	Primitive Horses
	65	Cretaceous Period	Last Dinosaurs
MESOZOIC ERA ("Middle Life")	140	Jurassic Period	Quarry Dinosaurs
	210	Triassic Period	First Dinosaurs
	245	Permian Period	Primitive Reptiles
	290	Pennsylvanian Period	Giant Insects
PALEOZOIC ERA ("Ancient Life")	320	Mississippian Period	Brachiopods
	360	Devonian Period	Primitive Fishes
	410	Silurian Period	"Sea Scorpions"
	440	Ordovician Period	Nautiloids
	500	Cambrian Period	Trilobites
	570		

Fossils older than Cambrian age are rare. This earlier span of time is usually called, simply, Precambrian.

205

Scientists have been aware through **fossil records** and historical records about when dinosaurs went extinct because - their fossils were found throughout the **Mesozoic era**, but were *not* located in the rock layers (strata) of the **Cenozoic era** - which starts about 65-million years ago and extends to the present (See the chart above). So, science knew that dinosaurs went extinct some *64-66 million years ago*, but that was all. Many different ideas and theories about how and why the dinosaurs were rendered extinct have been presented over the years.   A group of scientists back in 1980 proposed fresh ideas concerning the *mechanism* for the *"K-T extinction."* They're theory was that the presence of iridium in the layers could only have been caused by a meteorite.

Since their hypothesis was first proposed, **the search** for the *"an additional cause"* of the K-T extinction has been a busy area of scientific research. It incorporates scientists from many **different fields** including astrophysics, astronomy, geology, paleontology, ecology and geochemistry. Media coverage over the last 15 years has dwindled, while some paleontologists have since lost interest in the issue, preferring to study *how the dinosaurs and their contemporaries "lived" rather than why they died.*

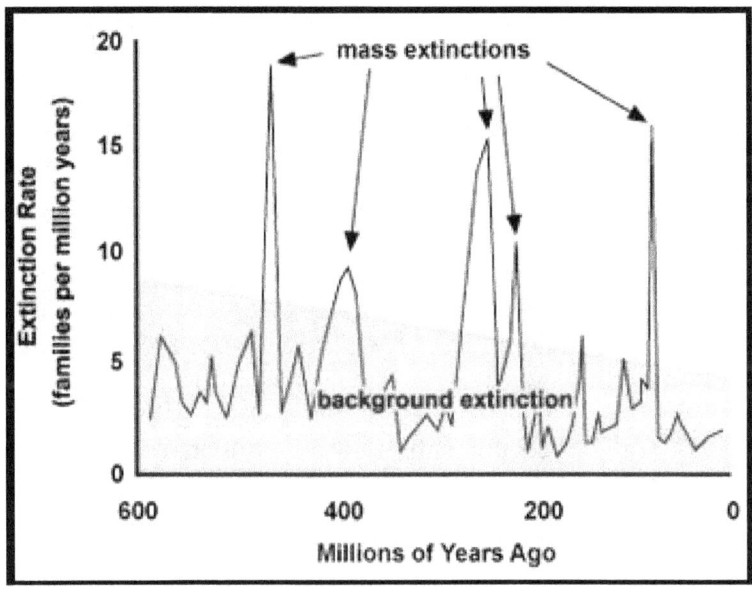

Mass Extinctions have occurred previously in Earth's history; however, the K-T extinction represents an example of when a mass extinction became a significant episode in evolutionary history where more than **50%** of all known species living at that time went extinct in a short period of time (less than 2 million years or so).

Catastrophic as that may be, we know of several mass extinctions in the history of Earth where the K-T extinction is not even the largest! The largest would be the "*Permo-Triassic*" extinction, between the Permian and Triassic periods, of the Paleozoic and Mesozoic eras.

In this yet another **catastrophic event**, life on Earth nearly was wiped out entirely - an estimated **90%** of all species living at that time were extinguished. Scientists are fairly sure that those extinctions were mainly due to many changing global conditions at that time.

Scientist Dr. Joseph kirschvink and colleagues onSeymour Island have begun to compare the K-T extinction to this event and discovered that about **60%** of all species that are present below the K-T boundary are not present above that line.

The Age of Dinosaurs

The Age of Mammals

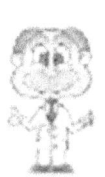

Large groups of organisms, including some members of Foraminifera, Echinodermata, Mollusca, and the marine Diapsida all were devastated by the K-T event. On land, the Dinosaurs of course went extinct, along with the Pterosauria. Mammals and most non-dinosaurian reptiles seemed to be relatively unaffected.

In fact, dinosaurs were not even among the most *numerous* of the casualties - the **worst** hit organisms were those in the **oceans.**

The terrestrial plants suffered to a large extent, except for the **ferns,**

which show an apparently dramatic increase in diversity at the K-T boundary, a phenomenon known as the fern spike.

A fresh group of Scientists freely admit that they have **no conclusive answer** to the mystery of the K-T event citing several complications: The first complication lies with the **fossil records themselves.** The Fossil Records at times are **incomplete** because the data has been found to be **inconsistent**.

The Paleontologists keep finding new fossils that are **hidden in the rocks.** Most data on the K-T event comes from North America, which is one of the few areas known that has a somewhat **continuous fossil record** (remember, fossils are **only** formed under certain rare conditions, and are only found in **sedimentary rocks** (see below)

The infamous **Hell's Creek** locality in Montana is one such continuous site enclosing the K-T boundary. UCMP researchers have led and continue to lead expeditions to Hell's Creek, gathering fossils from the rich fossil beds. Scientists now say that the secret to the K-T event may lie within these collections.

Additionally, scientists say we don't know much about what was occurring in the rest of the world at the time of the K-T event. The marine fossil record gives **substantial hints** about what was occurring within the sea, but fails to address how and what went on in the terrestrial realm.

The next complication is that *extinction itself* is a very complicated event; it is not simply the death of all representatives of a group. It is the cessation of the origination of new species that renders an entire group extinct; if species are constantly dying off and no new ones originate through the process of evolution, then that group will go extinct over time no matter *what* happens.

New dinosaur species **ceased to originate** around the K-T boundary; the question is, were they killed off (implying causation, especially a catastrophe), or were they not evolving and **simply fading away** (perhaps implying gradual environmental change)?

Yet more complications appear like *time resolution concerning the determination of* the age of rocks or fossils that are millions of years old is complicated and **carbon dating** only has a reasonable resolution when used with organic material that is less than about **50,000 years old**, so it is useless with the 65 million year old K-T material. Other methods of age determination are often even less accurate or less useful in certain situations.

So the truth is - scientists really don't know for sure exactly when or how the dinosaurs went extinct and matching events precisely to give a picture of what was happening at a specific moment in the Mesozoic is not easy. Thus, the ultimate question of a gradual decline of dinosaurs vs. a sudden cataclysm is **very difficult to answer** without the addition of a lot of good data.

To truly understand the situation of the dinosaurs around the K-T boundary, scientists are attempting to understand the **paleoecology** of that time on Earth. In this sense, they're using the term Paleoecology as an extension of the discipline of **ecology** or attempting to understand the interactions of organisms with their environment. Additionally, they are using both the geological paleontological data to tell them about the **abiotic** (non-living) environment) along with the paleontological (what plants and animals are found as fossils tell you a lot about the **biotic** (living) environment) evidence.

Important to note: With the problems of the fossil record and time resolution, it is difficult to understand the paleoecology of a region at a specific time in the past. However; a proposal by the current researchers involves a theory that helps them to understand the limitations of the fossil record. Their theory states that groups of organisms may *seem* to go extinct in the fossil record before they *actually do*; in other words that many times a false accuracy of the fossil timetable record prevailed rather than actual extinction. Thus, it is possible that some groups of organisms did not go extinct at the K-T boundary, and also possible that some organisms that seemed to have gone extinct earlier may have **survived** up to the boundary, and **then** gone extinct. This matter further complicates the important issue of the *selectivity* of the K-T extinction.

Science tell us that although many hypotheses about dinosaur extinction sound quite convincing and might even be correct – but it reminds us that a hypothesis is defined as an educated guess that sometimes means - maybe yes and maybe no. Some say it's not really science if they cannot be proven or disproved.

Even with the best hypothesis, such as the **impact hypothesis**, it is very difficult to prove or disprove whether the dinosaurs were rendered extinct by an event that occurred around the K-T boundary, or whether they were just weakened (or unaffected) by the event.

This is not to say that all extinction hypotheses are not science; many are excellent examples of good science, but a linkage of direct causation is a problem.

New insights about the asteroid thought to have killed off the dinosaurs suggest it may have just been the final blow, and that the reptiles were already suffering from an inconsistent climate prompted by **volcanic eruptions** long before the meteorite struck. The research, detailed in the February 8 issue of the *"journal Science,"* adds to the ongoing scientific debate over what exactly killed off the dinosaurs.

That debate, which once revolved around the question of whether the culprit was an asteroid or volcano-induced climate changes, has evolved to consider the possibility that perhaps **multiple** environmental factors were involved. Using a high-precision dating technique on tektites— pebble-sized rocks formed during meteorite impacts—from Haiti that were created during the event, the team concluded that the impact occurred 66,038,000 years ago— **slightly later** than previously thought.

When error limits are taken into account, the new date is the same as the date of the extinction, the team says, making the events simultaneous.

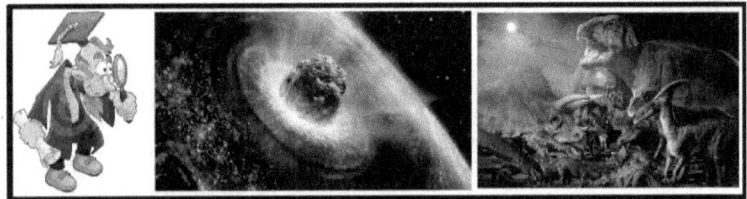

Additionally, scientists say the new findings should lay to rest any remaining doubts about whether an asteroid was a factor in the dinosaurs' demise.

"We have shown that these events are synchronous to within a gnat's eyebrow," **they said,** *"and therefore the impact clearly played a major role in extinctions."*

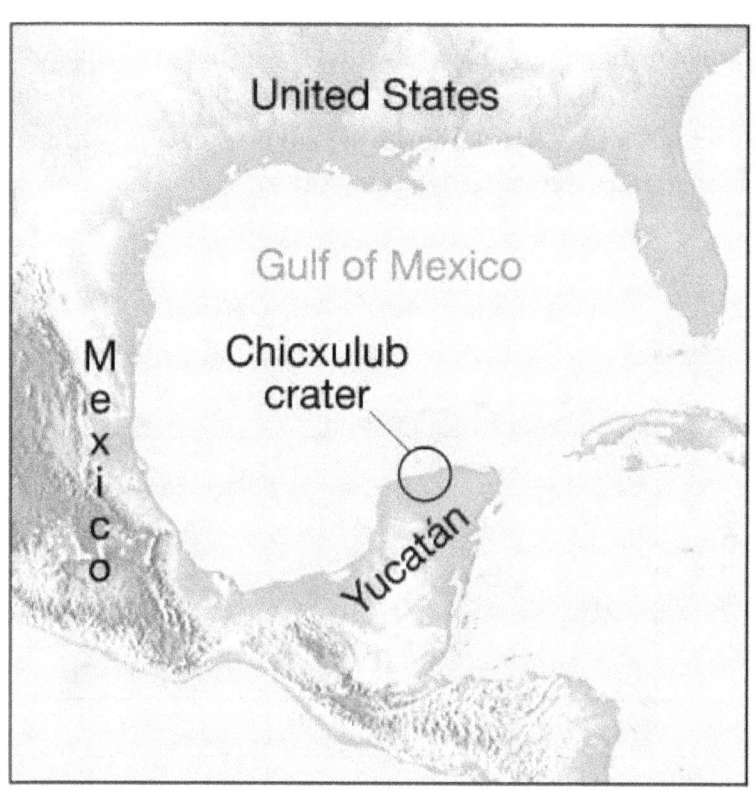

Chicxulub crater

That is not to say, however, that the asteroid—which carved out the so-called Chicxulub crater—was the **sole cause** of the dinosaurs' extinction. Evidence now suggests massive **volcanic eruptions** in India that predated the asteroid strike **also played a part**, triggering climate changes that were already killing off some dinosaur groups.

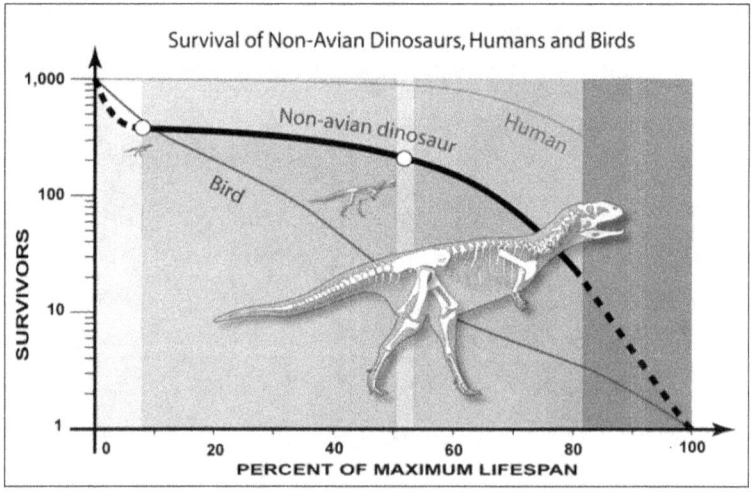

For example, nobody has ever found a non-avian dinosaur fossil exactly at the impact layer, hence, strictly speaking, the **non-avian "*dinosaurs*"** - those dinosaurs unrelated to birds - may have already gone extinct by the time of the impact.

The idea that volcanism was responsible for the dinosaurs' extinction actually predates the impact theory, and many of the other mass extinctions have been found to co-occur with large-scale volcanic eruptions.

Scientists including a physicist and planetary scientist, respectively, presented yet a *new* theory after discovering that a layer of clay that's found throughout the world and that coincided with the end of the Cretaceous period is enriched in *iridium.*

Iridium is an element rare on Earth but common in space rocks - they also proposed that a meteorite wiped out the dinosaurs.

As the impact theory took hold, especially with the more physical scientists ... the volcanists lost ground.

The impact theory gained even further momentum when another group of scientists re-discovered the 110-mile (180-kilometer) wide impact crater in the Yucatán Peninsula that dated to the boundary between the Cretaceous and Tertiary periods—the so-called KT boundary—when the crater's size indicated that whatever created it was roughly 6 miles (10 kilometers) in diameter. Although its presence was known by the Mexican government, for over ten years it was kept secret.

They figured that an asteroid of that size striking the Earth would have had devastating consequences, including destructive pressure waves, global wildfires, tsunamis, and a *"rain"* of molten rock reentering the atmosphere.

Additionally, they pointed out that much additional particulate matter would have stayed afloat in the atmosphere for weeks, months, perhaps years, blocking incoming solar radiation and thus killing plant life and causing catastrophic drops in temperatures

The once-abandoned volcanism theory has seen a revival of sorts in recent years, however, as a result of fresh insights about a period of sustained ancient volcanic activity in India and the discovery that dinosaur diversity may have already been declining before the asteroid strike.

The raging debate now is whether the Chicxulub impact was the *'smoking gun*,' as many researchers claim**,** *or* one of several causative factors. A popular theory is that a series of volcanic eruptions in India that produced ancient lava flows known as the **Deccan Traps** caused dramatic climate variations, including long cold snaps that may have already been cutting down the population of dinosaurs before the asteroid struck.

It seems clear that volcanism alone, if on a sufficiently massive and rapid scale, can trigger extinctions, a popular view now is that the impact was probably the final straw, but not the **sole** cause. The new volcanic theory still has some major questions it must answer however, like precisely how much the Indian volcanic eruptions affected the dinosaurs.

Mount Pinatubo Crater

Some people say if you look at the eruption of **Mount Pinatubo** (above) in 1991, it cooled the Earth for a short period of time due to the aerosol and the dust that was ejected, but, others say in the long run volcanoes probably pump more carbon dioxide into the atmosphere and actually warm the planet, at least temporarily.

It's also unclear how the **Deccan Traps eruptions** were spread out in time. We know that they started a few million years before the end of the Cretaceous and lasted for several million years after, extending even beyond the [asteroid impact], however, some people have suggested that there were clusters of eruptions that happened within a span of a few tens of thousands of years.

Knowing the timing of the eruptions is important, because if they were happening close to the end of the Cretaceous, it's more likely they played a role in the dinosaurs' extinction than if most of the eruptions happened two million years before. More precise dating of the volcanic ash layers in India could help answer some of the remaining questions. That's the next step of the puzzle.

Pinning down the cause of the dinosaurs' extinction isn't just of academic interest. It's important for us to understand how ecosystems respond to big disturbances. Whether it's gradual climate change or a catastrophic event, these are all things we have to think about as humans on the planet today.

Scientists remind us that Impacts by asteroids and comets have happened many times in the past, however; they also remind us that so far scientific evidence has concluded that only one mass extinction in history has ever been shown to be triggered by an impactor. In fact, the good news is that Scientists tell us that as the Universe continues to expand, Earth's orbit will see less and less asteroids and comets, and that every day the chance of mass extinctions from impacts becomes even more unlikely to happen in Earth's future.

It is my greatest hope that this book will help you answer some of those inevitable questions headed your way. The following chapter has been written entirely by my Granddaughters ages 7 and 12 and just like your own Grandchildren, I'm sure they have plenty to say and their own way of saying it. I find myself re-reading it and learning something new each time I do. After all, somebody I love wrote it.

Chapter Fourteen   Part 1

What if?

Top 20
children's
questions
which baffle
Grandparents
⇩

**JayLynn and Paula
Mother and Daughter**

Part 1 - *By Jaylynn Loreman*

This book is supposed to be about questions and answers so this chapter (14) has them, lots of them. Here to start with is a list of the top 20 questions from kids, followed by more questions including the really big one in Part 2 - "What is a Grandparent?" By Nancy Woodard

1. How is electricity made?

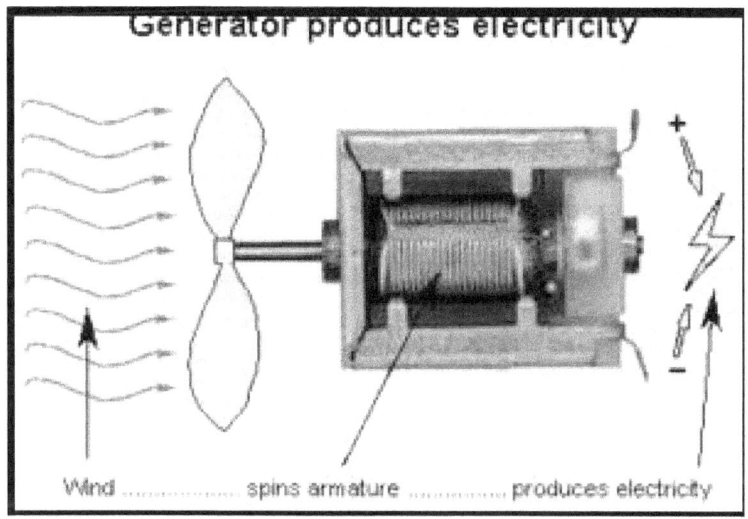

2. What are black holes?

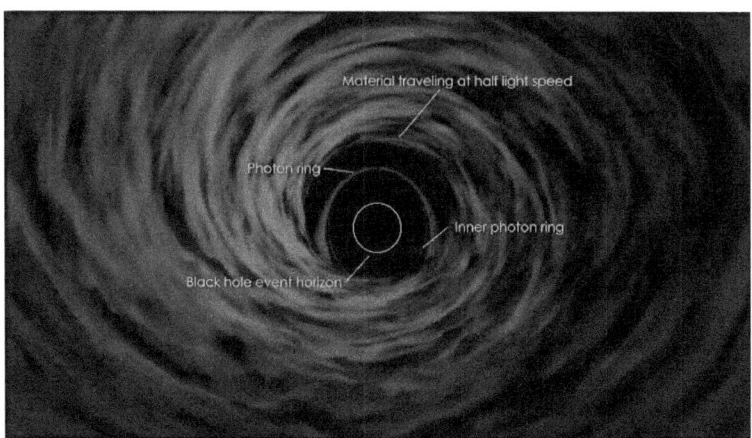

3. What is infinity?

Infinity never ends.

4. Why is the Sea blue?

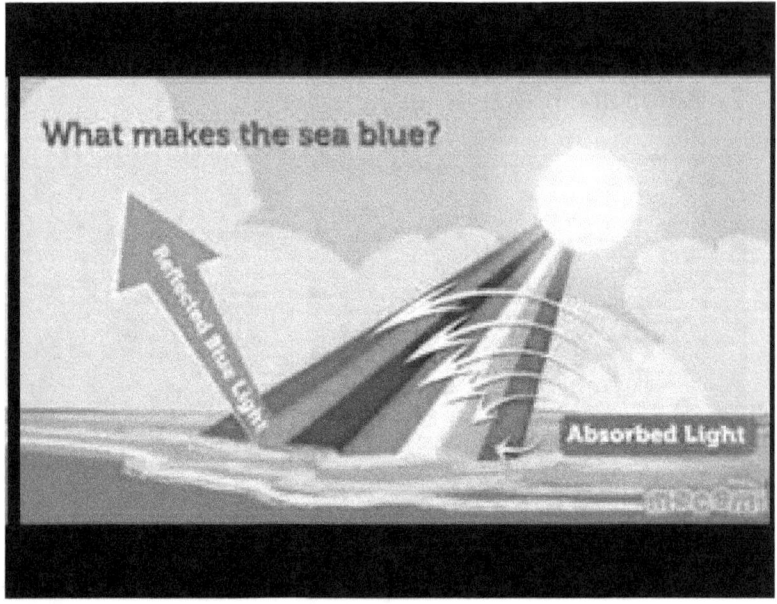

Reflection of blue light makes the Sea blue.

5. What is a leap year?

6. How do birds fly?

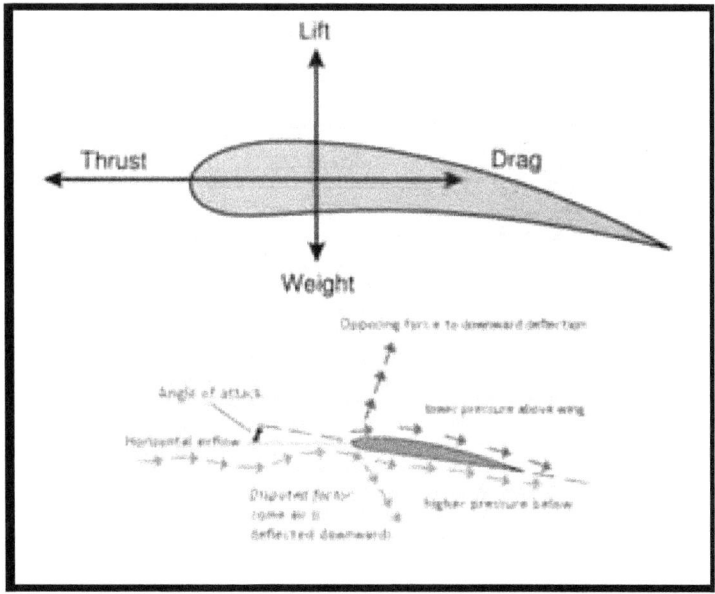

7. Why does cutting onions make you cry?

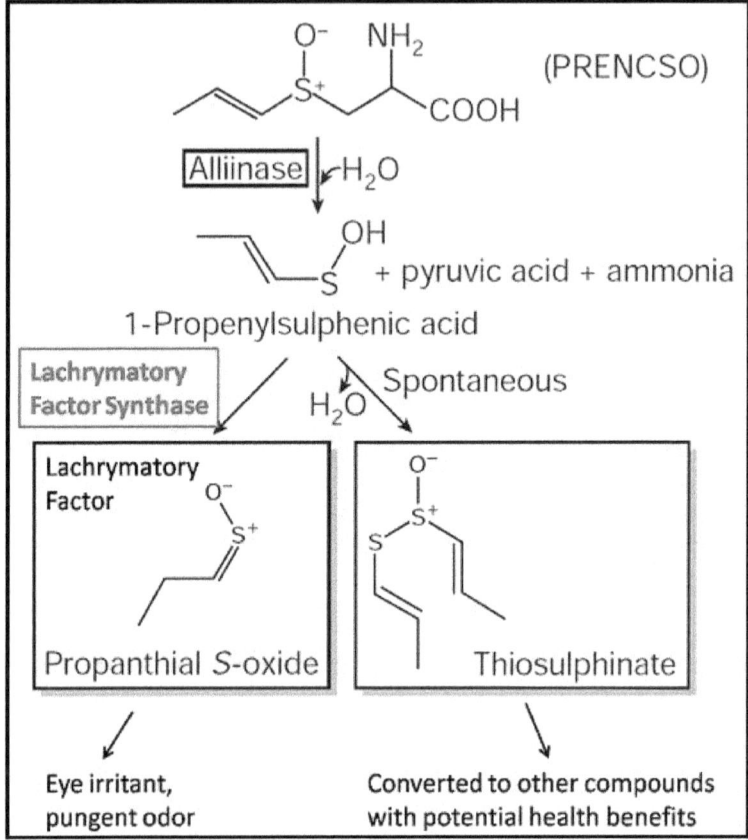

They give off eye irritating gas when we peel them that stinks and makes our eyes water.

8. Where does the wind come from?

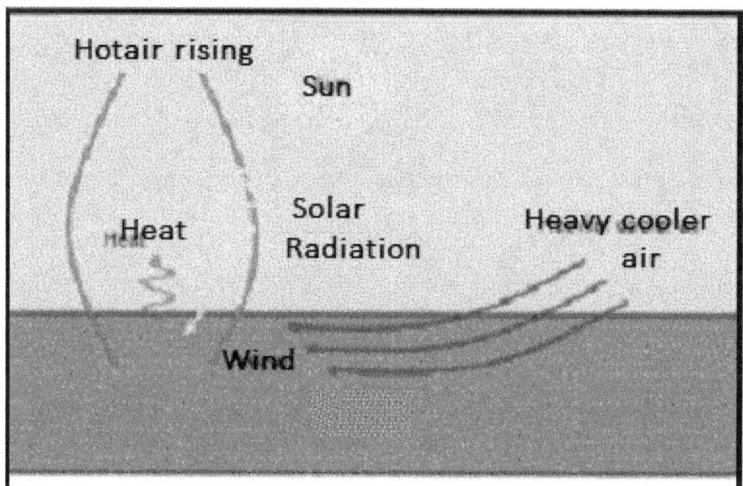

The solar radiation from the Sun heats the Earth. This heats the air, causing the lighter - hotter air to rise. Heavier, cooler air rushes in to take the place of the hot air rising - this causes wind.

9. Why is the sea salty?

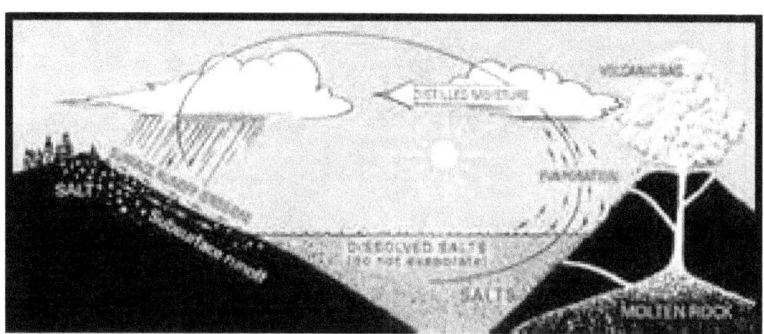

Because salt doesn't evaporate, it stays on the bottom.

10. How big is the world?

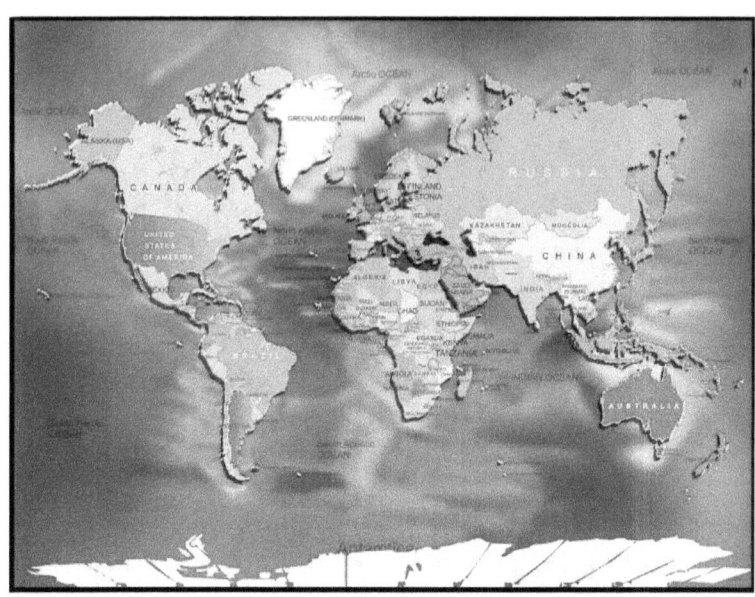

11. What happens to us when we die?

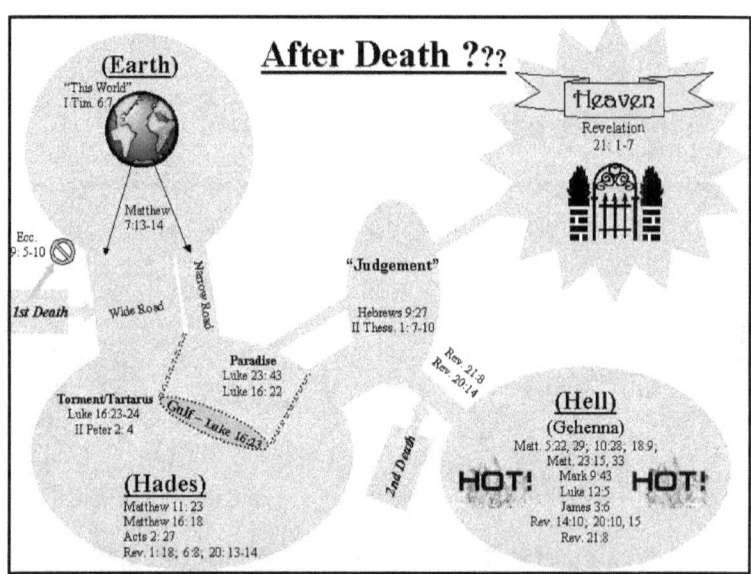

12. What is a prime number?

A Prime number is any number greater than 1 and has exactly 2 factors; 1 and itself.
Ex: 2, 3, 5, 7, 11...

13. Is God real?

God is to each person a unique feeling and idea. God is as unique to each of us as we are to each other.

14. What makes thunder?

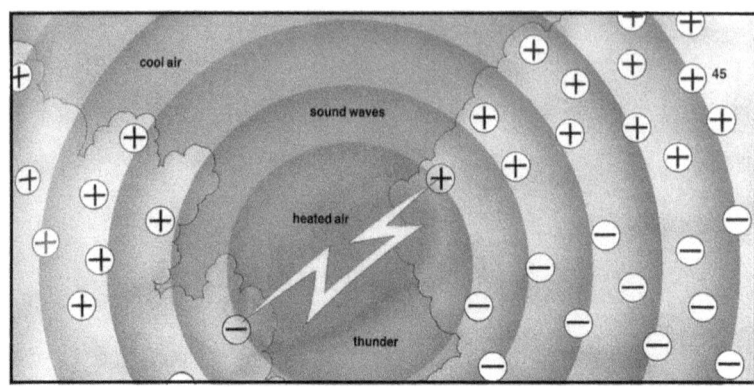

15. Why do you blink?

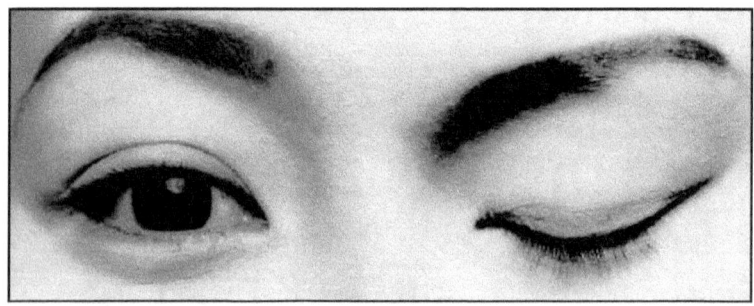

You blink to moisten and clean your eye.

16. Where do babies come from?

They come from fertilized eggs.

17. How do planes fly?

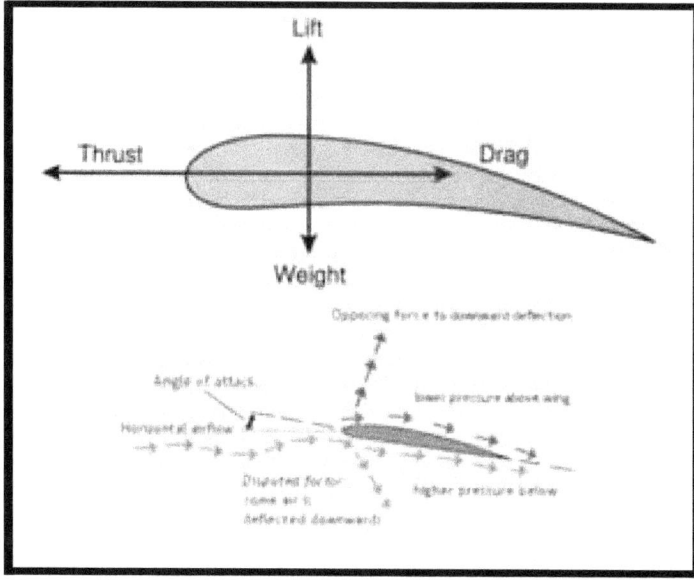

Planes fly for the same reason as birds.

18. What is time? Nancy: *"It's the stuff that makes clocks work."*

19. How does Santa Clause get down the chimney?

Magic.

20. Where does water come from?

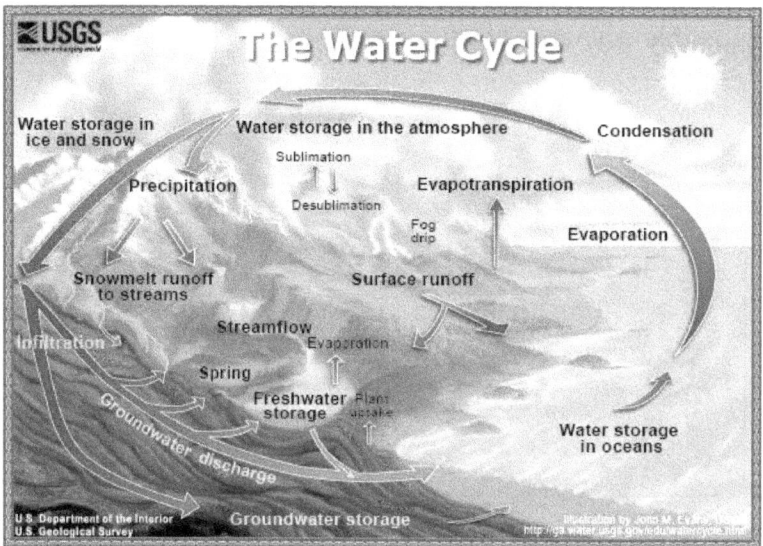

Where did people come from?

We can only imagine - but what if ………

Chapter One narrows it down pretty well that all of the people on Earth that are alive today came from the same small group of people that were left after the great human die off. Scientific evidence like geological records and DNA samples gives us an accurate explanation for that fact.

But one can't help but wonder about <u>who</u> were in that small group of people's ancestors?   Cavemen? Adam and Eve? The exact date the first humans appeared remains a mystery. There are libraries full of materials from all over the world filled with geological evidence like pottery, tools, weapons and even skeletons that evoke every possible scenario of man's early beginnings. In truth, we can't be sure. We can only imagine.....

Who's your Daddy?

Did we live in caves?

When did we learn to make fire?

Did we have pets?

Did we draw pictures ?

Did God make the Dinosaurs?

Did we have weapons and tools?

Was this what our neighborhood was like?

Is the story about Adam and Eve true?

How do we know what is true and what is not?

Did people and animals share with each other?

Where did people go to the bathroom ?

Why did we have a tree of life if it was bad?

Didn't the Bible say that man could eat anything that grew on Earth?

Where did Adam and Eve go when they got kicked out of Eden?

In the picture it looks like Eve is giving Adam an apple......did he ask for it?

Why don't the *Hebrew Scriptures* include the story of Noah's ark?

Is it one man's power or is it a miracle?

Why are some traditions hard to understand?

Albert Einstein always said, *"Until you can explain something simply — you haven't really learned it."* Good advice.

"The problem in society is not kids not knowing science. The problem is adults not knowing science. They outnumber kids 5 to 1, they wield power, they write legislation. When you have scientifically illiterate adults you have undermined the very fabric of what makes a nation wealthy and strong".
- Neil DeGrasse Tyson

Science that is based on facts can often be applied to more than one situation. It can be very useful indeed. Space science is more interesting to some than it is to others.

The Grandparent's Space Quiz
20 – Questions (Extra Credit)

Space is a mysterious place but there are plenty of things we are sure about when it comes to that vast space that surrounds us here on earth. Challenge what you think you know with our fun space quiz. Perfect answers that will amaze your Grand kids, this quiz will get you thinking about the fascinating topics of space and astronomy. Test your knowledge of planets, stars, moons, astronauts, our solar system, galaxy and more space related trivia. Try answering the twenty questions on your own.

1. What is the closest planet to the Sun?

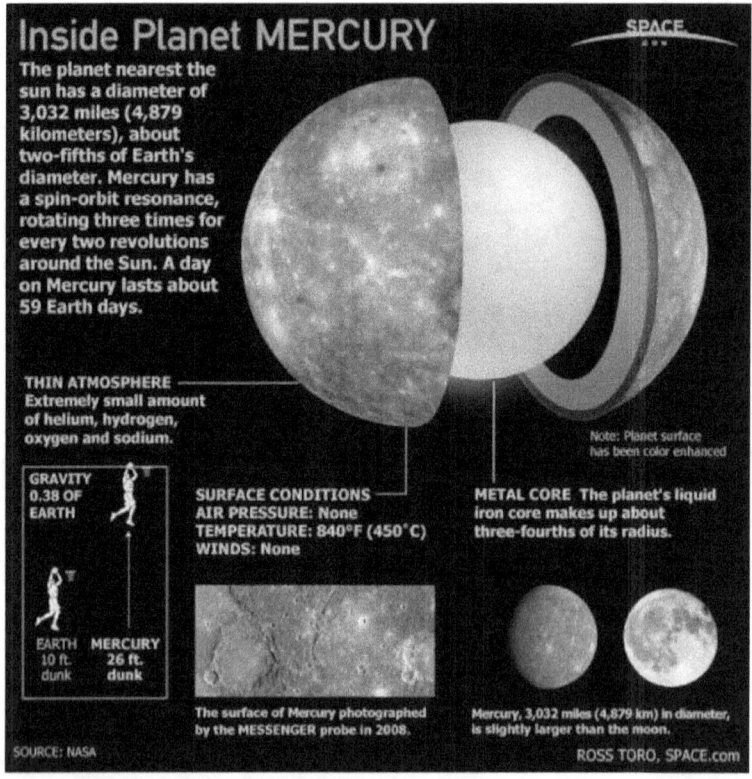

Inside Planet MERCURY

SPACE

The planet nearest the sun has a diameter of 3,032 miles (4,879 kilometers), about two-fifths of Earth's diameter. Mercury has a spin-orbit resonance, rotating three times for every two revolutions around the Sun. A day on Mercury lasts about 59 Earth days.

THIN ATMOSPHERE
Extremely small amount of helium, hydrogen, oxygen and sodium.

Note: Planet surface has been color enhanced

GRAVITY
0.38 OF EARTH

SURFACE CONDITIONS
AIR PRESSURE: None
TEMPERATURE: 840°F (450°C)
WINDS: None

METAL CORE The planet's liquid iron core makes up about three-fourths of its radius.

EARTH 10 ft. dunk   MERCURY 26 ft. dunk

The surface of Mercury photographed by the MESSENGER probe in 2008.

Mercury, 3,032 miles (4,879 km) in diameter, is slightly larger than the moon.

SOURCE: NASA

ROSS TORO, SPACE.com

Answer: <u>Mercury</u>

2. What is the name of the 2nd biggest planet in our solar system?

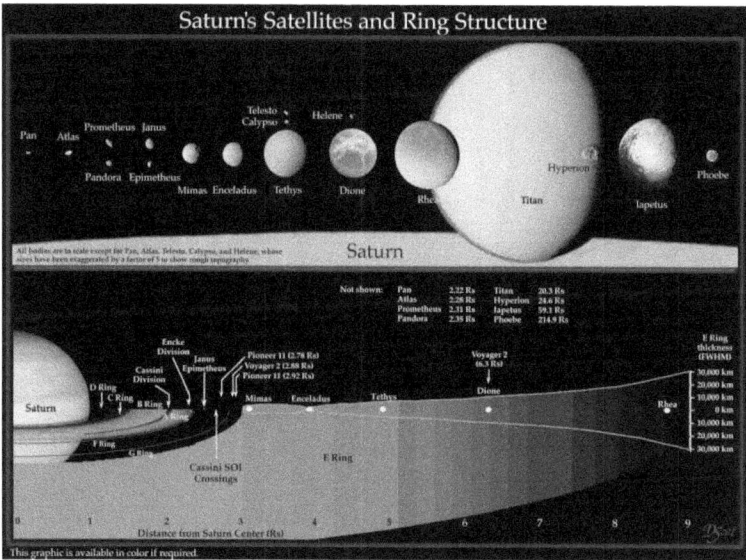

Answer: <u>Saturn</u>

3. What is the hottest planet in our solar system?

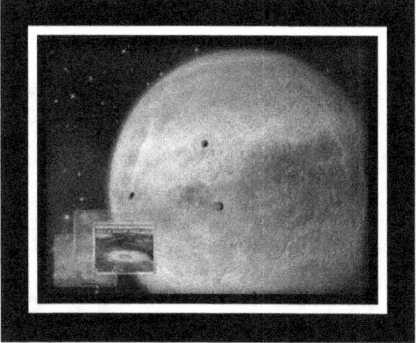

Answer: <u>Venus</u>

4. What planet is famous for its big red spot on it?

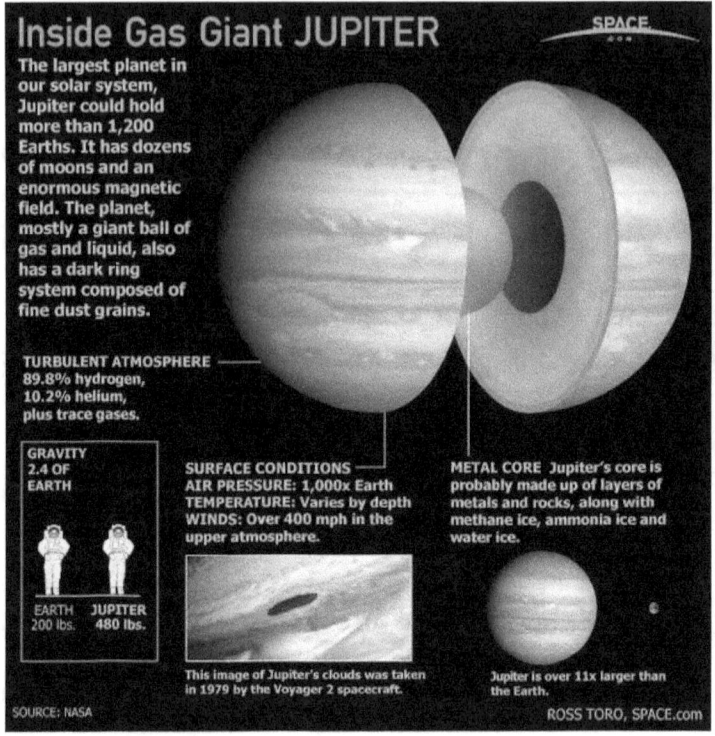

Answer: <u>Jupiter</u>

5. What planet is famous for the beautiful rings that surround it?

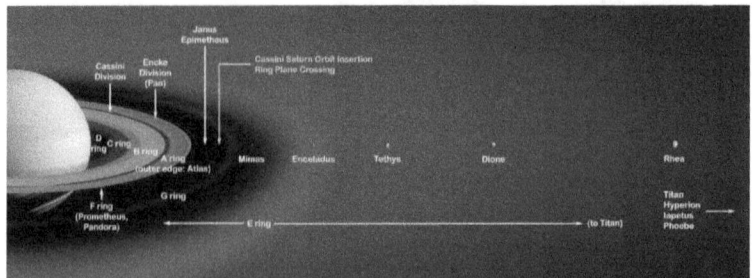

Answer: <u>Saturn</u>

6. Can humans breathe normally in space as they can on Earth? <u>*Answer: No*</u>

7. Is the sun a star or a planet? <u>Answer: a star</u>

8. Who was the first person to walk on the moon?

Answer: <u>Neil Armstrong</u>

9. What planet is known as the red planet?

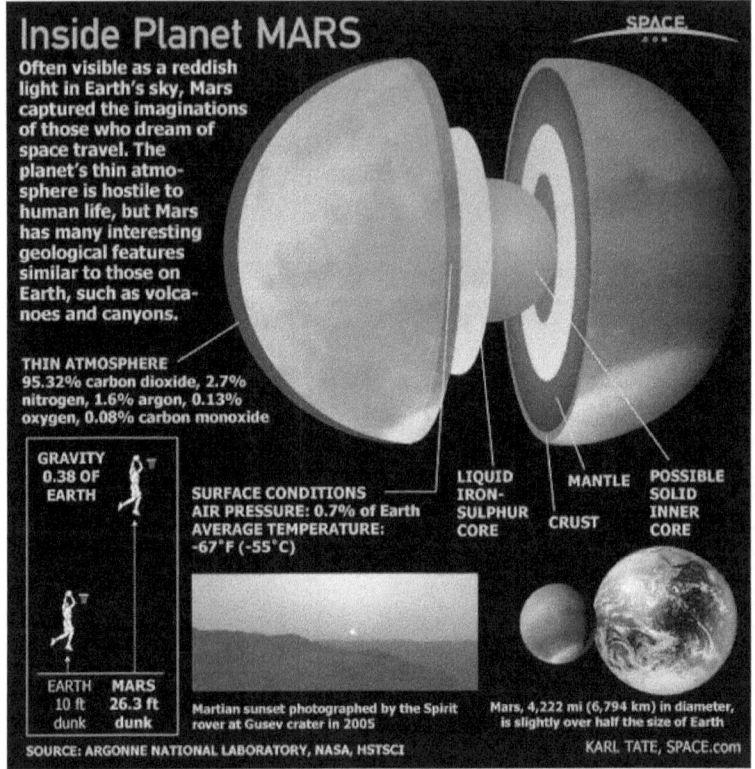

Answer: <u>Mars</u>

10. What is the name of the force holding us to the Earth?
Answer: <u>Gravity</u>

11. Have human beings ever set foot on Mars? Answer:
<u>No, not yet!</u>

12. What is the name of a place that uses telescopes and other scientific equipment to research space and astronomy?

Answer: **An Observatory**

13. What is the name of NASA's most famous space telescope?

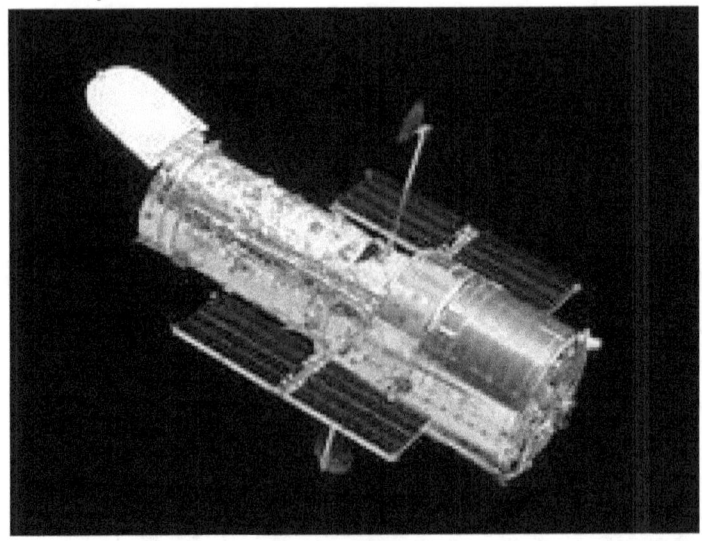

Answer: The Hubble telescope

14. Earth is located in which galaxy?

Answer: The Milky Way

15. What is the name of the first satellite sent into space?

Answer: <u>Sputnik</u>

16. Ganymede is a moon of which planet?

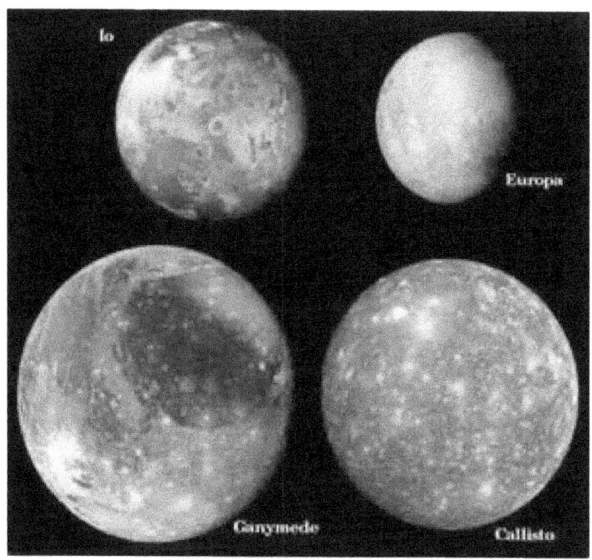

Answer: <u>Jupiter</u>

17. What is the name of Saturn's largest moon?

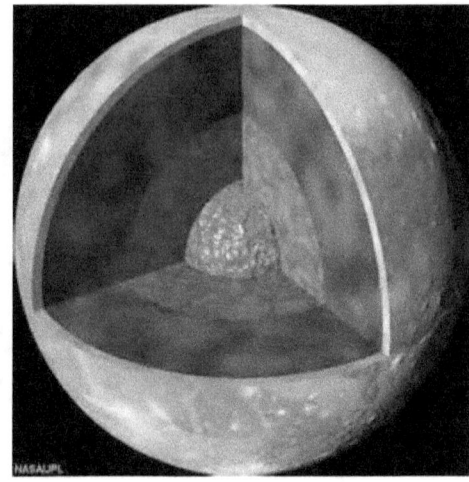

Answer: <u>Titan.</u>

18. Olympus Mons is a large volcanic mountain on which planet?

Answer: <u>Mars.</u>

19. Does the sun orbit the Earth?

Answer: <u>No</u>

<u>The Earth orbits the Sun.</u>

20. Is the planet Neptune bigger than Earth?

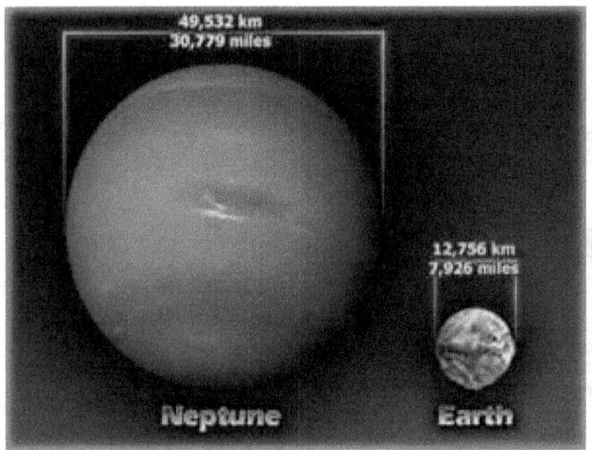

Answer: <u>Yes</u>

Part 2

What are
Grandparents?
by *Nancy*
Woodard

Nancy

Grandparents are:

Grandparents...

Are family historians
Believe in having fun
Cherish family get-togethers
Delight in telling stories
Encourage big dreams
Find ways to stay in touch
Give meaning to later life
Help in any way they can
Inspire wonder and mystery
Just want what's best for you
Keep family traditions alive
Love to takepictures
Matter more than words can say
Never run out of hugs
Open their hearts to you
Provide roots and wings
Quickly come if needed
Remember special occations
Share values and memories
Teach things no one else can
Unconditionally love you
Value spending time together
Write and call often
Xperience childhood again
Yearn for your happiness
Zest to makea difference

Why do Grandkids love their Grandparents?

25 Reasons Kids Love Grandparents

1. We're always up for an adventure. Whether it's a nature hike, a day of kayaking, or a weekend of camping, we like to get outside as much as the kids do.

2. We know lots of stories — and some of them are even true.

3. We can grow things. Maybe it's a garden of tomatoes and zucchini. Maybe it's just a potted plant. And if the kids help, we'll let them get their hands dirty.

4. We know our way around libraries and book stores. Where are the picture books? Where can kids find something for tomorrow's report on Conestoga wagons? We can be great guides.

5. We're not afraid to be silly. Did a grandchild make a mask or a funny hat? We know who will try it on. Is no one at home willing to listen to kids' knock-knock jokes? Guess whom they should call.

6. We let grandchildren take their time. When kids stay with us, they don't need to rush to get dressed in the morning. They know we will wait for them.

7. We make the best audience. When grandchildren learn a new poem, some fancy dance moves, or their first violin piece, they count on us to watch, listen — and applaud.

8. We have the coolest pets. Whether it's a cocker spaniel or a real, live, talking parrot, we share our animals with the kids. What's more, we let our grandchildren help take care of the animals.

9. We love to travel. We're up for a week in Yellowstone, a vacation at Disney, or a family cruise.

10. We knew our grandchildren's parents when they were kids, and we have the pictures to prove it. We can tell the kids stories about their parents that their moms and dads would never tell themselves.

11. We make great popcorn, not to mention brownies and sundaes. And we always let kids help, even if it gets messy.

12. We know tons of good songs, especially the ones kids learn in nursery school and kindergarten, and we love to sing along.

13. We're always thrilled to hear from our grandchildren, especially when they have something special to report.

14. We're walking history books. Most of us can remember when televisions had antennas, cars had fins, and phones had cords. When we take kids to a science or history museum, we make exhibits come to life.

15. We collect and display our grandchildren's art like museum curators, and we wouldn't trade a grandchild's first sunny-day scene for the Mona Lisa.

16. We send kids the coolest care packages when they go away to camp, to college, or just for the fun of it.

17. We pay our grandchildren to help out around the house, and sometimes we even beat minimum wage.

18. We believe whatever stories our grandchildren want us to believe, even when neither of us can stop laughing as the kids tell them.

19. We have skills to teach our grandchildren that they can't learn in school, like patching a tire, catching a fish, and sewing a hole in a favorite pair of jeans.

20. When our grandchildren aren't feeling well, we will snuggle up on the couch and watch cartoons with them, even SpongeBob.

21. We'll read our grandchildren's favorite books to them, over and over again. Then we'll go out and get new books that are even better.

22. Even if we live far away, we stay in touch and make sure grandchildren know that we're thinking about them.

23. We let our grandchildren teach us things, like how to use a cell phone and share photos online.

24. We want to make every day we spend with our grandchildren special. And when we're not with the kids, we go online to find more things to do with them.

25. We love our grandchildren unconditionally. Now, what could be better than that?

2017

Remember: *Time waits for nobody!*

Jaylynn Loreman   Nancy Woodard

Thanks, for sharing our book with us. For us it was great to spend time with our Grandpa, Dan Milburn, and of course with each other, but most of all, <u>with you.</u>

www.ingramcontent.com/pod-product-compliance
Lightning Source LLC
Chambersburg PA
CBHW071712170526
45165CB00005B/1979